Chasing Shadows

Investigating the Paranormal in

Illinois, Missouri and Iowa

LARRY WILSON

Huntzman Enterprises

Chasing Shadows - Copyright © 2011 by Larry Wilson

All rights reserved. This book, or parts thereof, may not be reproduced, distributed, or transmitted in any form, or by any means, or stored in a database or retrieval system, without prior written permission from the copyright owner, except for the inclusion of brief quotations in an article or review.

Cover Design Copyright © 2014 by Huntzman Enterprises

First Edition: September 2011

Second Edition: January 2015

ISBN-10: 0991129776

ISBN-13: 978-0-9911297-7-5

Published by Huntzman Enterprises

Printed in the United States of America

1 3 5 7 9 10 8 6 4 2

Acknowledgements

Thanks to everyone that made this book possible. A special thanks to my wife Kathy and my son Cory for their patience and support, as I traveled to secluded graveyards and haunted locations in the middle of the night conducting my investigations.

To Gary Hawkins my friend and mentor in the field of paranormal investigations who gave me guidance, encouragement and my first opportunity to investigate the supernatural.

To Paul Robinson, filmmaker and paranormal investigator, who gave me the opportunity to work with him investigating the paranormal and who taught me the art of making paranormal documentaries.

To my friends in the field of paranormal research who help make sense out of and offer reassurance to me in a field of endeavor that sometimes causes me to question my own sanity.

Books by Larry Wilson

CHASING SHADOWS

ECHOES FROM THE GRAVE

Table of Contents

Introduction   1

Why Ghost Hunting?   5

Chapter One: Anderson Cemetery (Graveyard X)   24

Chapter Two: Williamsburg Hill (Ridge Cemetery)   42

Chapter Three: The Legacy Theater   82

Chapter Four: Rockcliffe Mansion   97

Chapter Five: LaBinnah Bistro   124

Chapter Six: Kemper Military School   138

Chapter Seven: Morse Hill Hotel   159

Chapter Eight: Villisca Axe Murder House   179

Chapter Nine: Cumberland Sugar Creek Cemetery   208

Conclusion

Advice for Paranormal Investigators   230

The Supernatural   236

Where to Go Next   239

About the Author   245

INTRODUCTION

"When you have eliminated all which is impossible, then whatever remains, however improbable, must be the truth." - Sherlock Holmes (Sir Arthur Conan Doyle)

In the world in which we live, there are things that can be explained with science and logic. Then there are things beyond the realm of science and logic: things that go bump in the night, strange sounds, and ghostly images that occur without warning. For generations, these types of occurrences have been reported throughout the world. Every city and town has a place that everyone knows not to visit after the sun goes down, for fear of coming face to face with the beings that have made these stories so frightening.

The Midwest has its share of such places, and in this book we will explore some of them firsthand through investigations that I have conducted. We will explore haunted and mysterious places in Illinois, Missouri, and Iowa. Locations such as Illinois' haunted Ridge Cemetery at Williamsburg Hill, Springfield's Legacy Theatre, and the mysterious *Graveyard X*. We will explore Hannibal, Missouri's Rockcliffe Mansion, the historic Morse Mill Hotel in Morse Mill, Missouri, and Booneville, Missouri's Kemper Military School, which was the setting for the movie *Child's*

Play III. We will travel to Villisca, Iowa to investigate a place that has personally changed my life, the way that I view the supernatural, and the way that I approach paranormal investigating in general. These are all places where extraordinary things occur that cannot be explained by mere science and logic, things that I have personally seen and heard.

Are these unusual occurrences signs or evidence of ghosts and hauntings? If so, why are these particular locations haunted and not other, similar places? Do portals and windows exist? Or is the veil between the living and the dead somehow thinner at these locations, letting spirits travel between different dimensions and allowing us to experience extraordinary events? This book is my humble attempt at answering some of these questions.

What you are about to read is true. The stories in the following chapters describe a few of the most memorable paranormal investigations that I have conducted. The events as described are actual experiences and encounters, not simply legends or folklore. One particular location, as you will read, has completely changed the casual manner in which I once investigated the paranormal. Through these experiences, my perceptions of life, of ghosts, hauntings, and the supernatural, as well as what I have been taught to believe, have been changed forever.

It is my hope that those who read this book, especially those who plan to pursue paranormal research and investigation, will realize that ghost hunting is not a field of endeavor to be taken lightly. It is not fun and games or simply a matter of thrill seeking. This, my friends, is real!

I have never been a skeptic. As I believe that skeptics, are narrow-minded, and have already come to a conclusion, before they examine the evidence at hand. Instead, I describe myself as being open minded to all possibilities.

Long-time paranormal investigator, Gary Hawkins, first invited me to accompany him on an investigation to, as Gary put it, "see for myself if this stuff is real." At the time, I truly believed that I would go on a couple of investigations with him, nothing would happen, and that would be the end of it. Well, to put it simply, I was wrong! Eight years later, I am still at it, fully involved in paranormal investigation and research. Why? Because I have seen, heard, and felt things for which I have no explanation.

I will give you many examples of the types of things that I have experienced. So, for this reason, I continue on my journey and adventure to the land of the unknown, the paranormal, and the supernatural.

To those of you who have already set sail on your own adventure into the supernatural, who have had that unexplained encounter that keeps you thinking about what you experienced each and every day of your life, you already understand what I am talking about. However, to those of you who enjoy reading books about strange and unexplained encounters or enjoy watching one of the many TV shows on ghost and hauntings, but have not quite decided if you want to venture out into the world of the supernatural, my advice is to keep your distance and leave the door closed until you are positive that you are ready to open it. Because once you open the door to the unknown, you may not like what you find on the other side. Once that door is opened, it may be hard to close again.

WHY GHOST HUNTING?

The question, "Why ghost hunting?" is an important one. The answer will help paint a picture for you, which will make things quite clear and give you a better understanding of my need for this type of adventure.

You will see that there is really nothing else that would satisfy my need to know about supernatural and spiritual things, like paranormal investigation does. Once I got a taste of the supernatural, and it was proven to me beyond a reasonable doubt that this stuff is real, I was so intrigued by it that I am in this adventure for the long haul. Am I obsessed with the paranormal? Maybe I am, in some ways. I thirst for understanding and knowledge to discover what lies on the other side. If finding out that there may be more to life than what I have been told is an obsession, then I am okay with being classified as obsessed with the paranormal.

I have always been interested in mystery, intrigue, and adventure. As a child, the only books that my parents would find me reading were either related to baseball, my other passion, or books pertaining to ghosts, hauntings, and mysteries. The world of baseball is governed by rules and guidelines that are known. I am sure that the realm of the supernatural is also governed by such rules and guidelines, through some type of cosmic physics or intelligence, but so far the rules have not yet been discovered or explained to us.

It does not appear to me that mainstream science plans to pursue, or even admit, that something else exists. Something, that may come from other realms, or plains of existence, which seems to somehow be able to interact with our three dimensional world. Mainstream science may have a lot to lose if they pursue answers to supernatural questions, but they have much more to gain if they unlock the secrets to the other side.

Therefore, in order to discover the rules of the supernatural, it will take paranormal investigators and fringe scientists, like myself, who are willing to think outside the box in order to come up with provable laws of supernatural physics.

My first encounter with the paranormal was on September 8th, 1966. The date is easy to remember, because the night of my experience was the debut of a cool new TV series, *Star Trek*. It was around 8:30 p.m. At that time we were living in the Petersburg area of Illinois. I remember that my mom was in the kitchen and my brother was doing homework. My dad was visiting at a friend's house.

Our living room was set up with our sofa under a window. It faced the north and rested up against the wall.

Being bored, I decided to look out the window. I climbed up on the sofa and knelt on the seat, my chin resting on my interwoven fingers. As a child of seven, I enjoyed looking out this window at night, to watch trains that would stop at a small depot that was located on a hillside, about a mile north of our home.

It was dark. I unclasped my hands, cupping them to my face in order to block the glare from the living room light that was reflecting on the window. I pressed my nose against the glass in order to see outside. I remember this night as though it was yesterday. As soon as I looked out of the window I could see the stars, as it was a clear night. Immediately, I noticed how large the moon looked. It appeared to be the biggest and brightest moon that I had ever seen. The color was a strange rich and vibrant orange, and its shape was perfectly round. I looked at the bright orange moon for a couple of minutes. I was completely fascinated by it. Then something caught my attention. I glanced to my left, looking higher up in the night sky, and saw the bright yellow, crescent shaped moon! I quickly looked back at the orange object, which I had originally thought was the moon, and then looked back at the actual moon. The moon appeared to be much smaller than the other object. One thing that I will never forget is how perfectly round the object appeared to be.

I continued staring at it, as I could not take my eyes off of it. Then, all of a sudden, it appeared to accelerate at a high rate of speed before coming to an abrupt halt. It now appeared to be half the size that it had been before. A moment later it shot off again at a high rate of speed and again came to a sudden stop. It repeated this behavior one more time, before it finally accelerated and disappeared into the northern sky.

Over the course of the next few weeks something strange began to happen. I would think of something and the next thing I knew, it would occur. It was not something that happened all of the time, but generally when I would get what I would describe as an *eerie* feeling. A thought would come to me and the next thing I knew it would manifest itself into reality.

One example of this was a day I had a premonition that someone was hit by a train. Later that evening I was laying on my parent's bed, staring out the window, when I observed a car driving through the railroad crossing that was across the street from our house. As I watched, a train struck the rear end of the car. The woman driver sustained injuries that resulted in her being taken to the hospital.

On another occasion, I was in school and we heard the village fire alarm. I looked over at one of my classmates and had a premonition that it was his house that was on fire. I recall hearing him jokingly say, "I sure hope it's not my house." Later on, family members came to the school to pick-up him up because it had, in fact, been his house that had caught fire.

After about two months, these *visions* stopped. But even as a child, I recognized what was happening to me. I truly believe that these premonitions were enhanced for a short time by what I saw that night. It is my hope, that someone who reads this book, and who lived in the Petersburg area at the time, may have also witnessed the same event. If so, I wonder if they experienced premonitions like I did.

In the summer of 2010 I was attending a paranormal consortium in Springfield, Illinois. I was explaining what I had seen as a child, back in 1966, to the Illinois Mutual UFO Network (MUFON) director, Sam Maranto. During our conversation, Sam would nod in an affirmative way, as if to say, "Yes, I have heard a similar story before." After I finished telling my story, Sam said, "Larry, did you ever think that when you saw the object accelerating at a high rate of speed, that maybe it wasn't traveling like you thought? What if I told you, that I believed you witnessed a wormhole in space, some type of inter-dimensional window or portal? Instead of it moving in the manner that you thought, what if when you saw the object getting smaller, you were actually seeing the wormhole closing up?"

What Sam said made complete sense to me, and the more I thought about it, the more I agreed with his theory.

Because of this childhood experience, and my interest in intrigue and adventure, I pursued a career as a private investigator. The ten years that I worked as a P.I. has been highly valuable to me as a paranormal investigator. As an investigator, you are trying to solve some type of a mystery. To do this, you have to think outside of the box. You come up with new ways and innovative ideas to solve a case. You are also required to conduct surveillances. Researching the paranormal is really no different. Except that, sometimes simply thinking outside the box is not enough. As I have found out, you sometimes, have to rebuild the entire box.

If you try to make sense out of things that do not seem to fit with the laws and rules of our three dimensional world, you won't. We have to make sense out of what we encounter by opening our minds and coming up with new ideas that do seem to fit. These ideas should help explain the encounters that we have, as well as how they manifest or come to exist in our world.

Although private and paranormal investigating have their similarities, I will say that staking out the dead in a cemetery, or in a haunted building, is quite different than staking out a living subject, in which you can actually see and evaluate with your own eyes. The use of our five senses:

touch, smell, hearing, seeing, and, occasionally, even taste, comes into play in both fields. However, our sixth sense, extrasensory perception (ESP), seems to kick in when things we cannot see are present. An example of ESP would be that icy chill on a warm summer's night, which causes the hair to stand up on the back of your neck or the feeling of being watched or followed when no one is around. Then, when you're reviewing your evidence, a shadow or mist appears in a photo that you cannot explain, or a voice is recorded that is not the voice of a member of your investigative team. You cannot ignore these types of occurrences.

So, the transition from private investigator to paranormal investigator is probably fairly easy to understand and helps explain why I have become so obsessed with investigating the supernatural. After all, what is more *mysterious* than disembodied voices and phantom people?

My interest in the paranormal and hauntings has grown even more intense in recent years after watching shows on TV such as *Unsolved Mysteries* and the Discovery Channel's *A Haunting*.

I finally decided to start pursuing investigations of the paranormal after a couple of my co-workers took haunted tours in Alton, Illinois. A few of them showed me some of the photos that they took while on their tour. Several oddities, such as streaks of light and strange mist, piqued my interest.

Enough so, that I decided to take a trip down to Alton to see for myself whether that *stuff* was real. I booked a private tour in Alton with paranormal investigator, Dr. Gary Hawkins, who also worked as a tour guide for a company called *Antoinette's Haunted History Tours*.

On my first tour with Gary, I invited two very close friends of mine and we arranged to meet him in the parking lot of Alton's *Information and Tourist Center*. Gary and I immediately hit it off and the three of us were given a tour of approximately twenty purportedly haunted locations in Alton, along with a haunted city park and two cemeteries. In the late 1990s, Alton was rated by *Fate Magazine*, one of the premier paranormal publications at that time, as the most haunted small city in the United States. Gary's knowledge of the haunted history, and the history of Alton itself, was quite remarkable.

Nothing really spectacular happened that night, other than in one of the cemeteries that we toured. I walked into a cold spot that felt twenty or so degrees colder than the seventy-two degree evening air temperature. I also had a feeling in this spot that something was not normal. This seemed to cause the hair on my left arm to stand up and goose bumps to run down the left side of my body.

I had such a great experience on the tour and was so fascinated with the haunted history of Alton that I decided to take a second tour a few weeks later. This time, I scheduled the tour with a group of friends from work, which also included my wife Kathy, who is not as thrilled with ghosts and the paranormal as I am. She always instructs me before I go out on investigations not to bring anything, meaning a ghost, home with me. I had always told her, "*Ghosts don't follow people home, dear.*" However, as you will see in a later chapter on the Villisca Axe Murder House, they can and do follow you home.

The second haunted Alton tour was scheduled for October, just a week before Halloween. The group consisted of my wife and I, along with three of my co-workers, as well as our guide, Gary Hawkins. We again met Gary at the *Alton Information and Tourist Center* around 7:00 p.m. on a rainy evening with a temperature of fifty-five degrees.

Gary took us on a trial run of his new route for the Halloween season, which was a haunted Civil War tour. During the Civil War, Alton was the location of a federal military hospital and prison used to house Confederate prisoners. In addition, there was a place called *Smallpox Island*, which is located on the Mississippi River and was used as a burial site for the Confederate prisoners who had contracted smallpox and died while incarcerated at the prison. He also took us to the Alton Cemetery, which has

13

many graves dating back to the early 1800s, along with the graves of Union soldiers from the Civil War.

Alton Cemetery is unique, in that, while the majority of the cemetery is city owned, part of it is a state cemetery. The state portion of the cemetery is the location of a ninety-three foot monument, erected in honor of abolitionist Elijah Lovejoy. Lovejoy was born in Albion, Maine on November 9th, 1802. In 1826, he came to St. Louis as a schoolteacher. He was ordained as a Presbyterian minister in 1834 and later published a religious newspaper, the *St. Louis Observer*. Lovejoy began to advocate the abolition of slavery and in doing so he enraged some of Alton's citizens by actively supporting the organization of the Anti-Slavery Society of Illinois. He then established the *Alton Observer* as an abolitionist newspaper. He continued publishing even after three printing presses had been destroyed and thrown into the Mississippi River. On the night of November 7th, 1837, a group of twenty supporters joined him at the Godfrey & Gilman warehouse, to guard a new press until it could be installed at the *Observer*. A pro-slavery mob assembled outside the warehouse where Alton's mayor tried to persuade the defenders inside to abandon the press. Lovejoy was killed by a shotgun blast while defending the warehouse.

The monument to Lovejoy was erected in the 1890s and is topped by a seventeen foot winged statue of Victory. It is *guarded* by; two thirty foot high, granite sentinel columns, which are mounted by, bronze eagles. Just behind the monument is a black gated, wrought iron fence, approximately six feet high, which separates the city cemetery from the state cemetery.

During the September tour, Gary pointed out a much smaller monument in the city cemetery. He went on to describe how a few years earlier, a photographer on one of his tours, had taken a group photo, which consisted of about twenty people. When the picture was developed there was an apparition standing near the small monument. During the tour, I viewed this photo on a website Gary had at the time. What I saw appeared to be a white figure on a horse next to the people standing in the photo.

As our tour group moved behind the Lovejoy monument, I noticed that it felt extremely cold. On our way over to the cemetery, I had seen a time and temperature sign at a local gas and convenience store. I recalled that the temperature was fifty-five degrees. The area in which we were standing had more of a feel of a temperature in the thirty-degree range. I asked if anyone else had noticed that the temperature felt colder in the area in which we were standing, as opposed to the rest of the cemetery. Everyone agreed that it did indeed feel colder. Lilia, one of my friends in the group, had a digital camera with her that was also capable of taking a small amount of video footage. She commented that her batteries had just gone dead and she had replaced them only a few minutes earlier.

One of the theories about ghosts is that when they are present they need energy to manifest, and one source of that energy can be batteries from cameras or other electronic devices. Some believe that spirits can also drain energy from people.

The cold spot seemed to take in an area of about ten feet by ten feet. When Lilia walked a short distance away and out of the cold spot, within minutes the batteries in her camera suddenly, and inexplicably, recharged.

I pointed to a small monument in the city cemetery. I asked Gary if this was the same monument that his photographer had recorded the image of the apparition that was on his website. Gary answered in the affirmative and began to explain where the tour group was standing in correlation to where the photo was taken.

Lilia began to film a short video of the spot where the photo of the apparition was taken. A few moments later, she reviewed the footage and exclaimed excitedly, "I think I've got something here!" She had recorded what appeared to be a ghostly image, visible from the neck down to just below the knees, walking through the cemetery. The image was hard to make out since the LCD screen on the camera was very small.

The following Monday, Lilia brought the memory stick from her camera to my home in Taylorville, Illinois. She had recently moved and did not have her home computer set up yet, so she had not viewed the video other than on her camera's LCD screen.

I downloaded the file to my computer. The length of the video was approximately twenty seconds and, to my disappointment, the only image visible was at about the ten second mark. At that point in the clip, a red and green sphere, or ball of light, floated erratically to about the center of the screen, where it stopped. The ball of floating light was impressive, as we had not seen it with our naked eyes, but I was still disappointed because the full apparition could not be seen on the video.

Later that evening, my son Cory came home and copied the file to a CD. He then played the CD on his laptop. He discovered that the ghostly image was now visible with the settings and type of screen that his laptop had.

What could be seen on Cory's screen was what appeared to be a soldier from the civil war era, complete with stripes on his sleeve indicating either a corporal or a sergeant. There was also some type of backpack or possibly a neckerchief around the neck. I also observed a shiny object directly to the side of the apparition, which we have theorized was possibly a saber. Several other shiny and pointed, yet indistinguishable, objects were moving in front of the apparition. This made us feel that there may have been a group of ghostly figures together and that those objects may have been the metal bayonets on the soldier's rifles.

Needless to say, with my friend capturing a ghostly image on video, during only my second time out in a cemetery at night, my doubts about the possible existence of ghosts and ghostly phenomena quickly disappeared. It made me believe that yes, there is something to all of this paranormal stuff.

The second experience that I had, that made me a believer, took place on a January night at Wolf Creek Cemetery, located near Spaulding, Illinois.

I had recently joined a local paranormal group in the Springfield, Illinois area. One of the members, Debbie, invited me to take an excursion to what is considered to be a very active cemetery. We set out about 9:00 p.m. from the group member's home and arrived at Wolf Creek about fifteen minutes later. The overall weather conditions were good, with a clear sky, and an almost full moon shining above. It was a very still night, with no wind at all and a temperature of around thirty-two degrees.

Wolf Creek is a very small cemetery located near the small village of Spaulding, Illinois; I estimated the cemetery to be maybe two or three acres in size. It has a gravel road that splits the cemetery in half. We drove into the cemetery and parked about halfway down the gravel road. We each had a digital camera, and Debbie had a cassette recorder, while I carried a digital voice recorder. Our plan was to walk around and take random photos to see if we could catch anything interesting on film. Debbie placed her tape recorder on a tombstone and I place my digital recorder on the ground next to a grave. The idea of the voice recording equipment was to try to capture possible EVPs.

EVP stands for Electronic Voice Phenomena. Paranormal investigators and researchers believe that voices captured on tape or digitally recorded are voices of the dead that are trying to communicate with us. The mysterious voices are not heard at the time of recording; it is only when the tape is played back that the voices are heard. Sometimes, amplification and noise filtering are required to hear the voices. My particular belief, based on the voices I have recorded, is that some of these voices seem to fit the mold of being ghostly, based on dialog which is recorded and ties in with the history of a location. Other voices that I have recorded seem to not fit in at all with the location that I was investigating. Sometimes, they seem to have a connection to the investigator.

A prime example of this was an investigation that Jay Nandi and I conducted at the Legacy Theatre in Springfield, Illinois. Even though Jay's heritage is from India, he does not speak

Hindi or have an accent because he was born and raised in the United States. But during the investigation, we recorded an EVP that had a strong Indian accent. It is very possible that the voice was a spirit, but I also believe that some of the voices that we recorded may have been from other realms, dimensions, or times. I also believe that sometimes we may even be recording our own thoughts. Someday, I hope to have a large enough sampling of EVPs to test some of my theories. At this point, I have over 200 quality EVPs from various locations.

Some EVPs are more easily heard and understood than others. EVPs vary in gender, age, tone, and emotion. They usually speak in single-words, phrases, or short sentences. Occasionally, grunts, groans, growling and other vocal noises are recorded.

Back at Wolf Creek, we walked around the cemetery for about forty-five minutes and randomly snapped pictures. Previous investigators, who had been to the cemetery on several occasions, had mentioned that it was a very active place and that they had recorded EVPs. Debbie pointed over to the tree line at the edge of the cemetery. She stated that in the summertime, she had witnessed colored balls of light in the trees. She had no explanation or theories as to what caused the lights, but there they were.

After about thirty minutes, something happened that still sends chills down my spine.

We were standing in the middle of the gravel road that split the cemetery in half. It was a clear, moonlit night, with no wind. The nearest tombstones were rather close to the ground, so we had an unobstructed view for about twenty yards directly behind us. There were two houses near the cemetery. The closest was about one-eighth to one quarter of a mile away. We were taking random photos facing the tree line, when all of a sudden we could hear someone running up behind us. My first thought was that one of the neighbors had seen the flashes from our cameras and had come over to chase us out of the cemetery. At the same moment, Debbie and I whirled around to see who was behind us, but no one was there. She looked at me and said, "I heard it too." Immediately, a cold chill ran down my spine. It was a creepy, but awesome feeling. The footsteps were so loud that I thought someone the size of a football player, like Dick Butkis, was going to run into us and knock us down.

I had just had my first encounter with what I believe was a ghost.

During the investigation, I snapped several photos. When I reviewed my film, several orbs appeared in the photos. Orbs, however, are not conclusive evidence of anything. There are many logical explanations for orbs in photos, including dust particles, moisture, and insects; however, over the years I have taken a handful of photos with orbs in them that cannot be explained. For example, last fall I took an exterior photo of an upstairs window at the Villisca Axe murder house. In the photo, you can see a very large cream-colored orb that appears to be solid. What makes this photo interesting is that, on the exterior facing side of the window, there is a white drip of paint that was left after the eaves of the house were painted. This paint drip is visible in front of the window, indicating that the paint is *between* the lens of the camera and the orb. So the orb is not simply a speck of dust or moisture near the camera lens, it is *behind* the glass and *inside* the house.

A Close up of the Orb in the Window

When I returned home from the Wolf Creek investigation, I reviewed the audio files on my digital recorder to see if I had captured any EVPs. At several places throughout the recordings I could hear what sounded like either a whispery breath or a gust of wind, but nothing really distinguishable. However, twenty-seven minutes and forty-four seconds into the recording, a loud and distinct horn could be heard. It was not a modern sounding horn, but the sound of an old steamboat or foghorn of some type. Neither Debbie, nor I, had heard this horn during our time at the cemetery. It was not from any passing cars or vehicles, because only three cars passed by the entire time we were at Wolf Creek and none of the cars sounded their horns.

My sense of adventure and investigative nature had originally drawn me to looking into the paranormal. But it was the events at Wolf Creek, coupled with the video recording of the apparition at the Alton Cemetery, which convinced me that, yes, there just may be something to all the tales that I had read, and the TV shows that I had watched throughout the years, alleging experiences with ghosts and the supernatural. Now, I had my own experiences and reasons to further delve into the supernatural.

Anderson Cemetery (Graveyard X)

CHAPTER ONE

Anderson Cemetery (*Graveyard X*)

Palmer, Illinois

Located not far from the small town of Palmer, Illinois is one of the most mysterious locations in Illinois. Anderson Cemetery, or as many prefer to call it: *Graveyard X,* is widely discussed by ghost hunters and paranormal investigators alike.

For years, there have been tales of unexplained lights in the cemetery. Some of these have been described as orange balls of light, and have been recorded on video, both during the daytime and at night. Others tell stories of hearing the sounds and voices of small children playing in the cemetery. However, there are no homes located nearby to account for these sounds. Apparitions appear in photos and digital thermometers record icy cold spots with no scientific or logical explanations for the temperature fluctuations. I have personally experienced these extreme temperature variations.

In order to keep the location from others, various books and internet postings have labeled it Graveyard X, leading to the perception that Anderson Cemetery is some kind of top secret place. All the stories and legends about the mysterious Graveyard X, however, piqued my interest enough to compel me to conduct my own investigation at this location.

My first challenge was finding the elusive Graveyard X, because all I knew about the whereabouts of this mysterious location was that it was in a secluded place, in rural, central Illinois. It seemed that other ghost hunters who had investigated Anderson Cemetery were not anxious to reveal its location. I do not think the reason for this secrecy was to prevent other legitimate paranormal investigators from investigating the cemetery, but to prevent those who may only be looking for a thrill and to possibly destroy, desecrate or litter the site.

I uncovered a paranormal message board on the internet in which someone was requesting directions to this particular cemetery. I located a response, which listed the latitude and longitude of an Anderson Cemetery in central Illinois. I plugged in the coordinates on *Google* and it took me to a site that listed cemetery locations. Unfortunately, these coordinates turned out to be for a *different* Anderson Cemetery. Coincidentally, both cemeteries are located in the same county, not far from each other. Using the same website, I was able to get directions to the correct Anderson Cemetery.

As it turns out, Anderson Cemetery, or Graveyard X, is located in rural Christian County only nine and a half miles from my home in Taylorville. So there it was, right under my nose all the time and I did not even know it. Oh well, so much for the intuition of a former private investigator!

I decided to make my initial trip to Graveyard X during the daytime in order to make it easier to find the location and map out the area. I would then come back at a later date to conduct my own paranormal investigation at night.

Sunday, March 4th, was a clear, sunny day, with a temperature around thirty degrees and a light breeze. So it was a good day to check things out. To my surprise, it turned out that Graveyard X was not a creepy, eerie place. At least not during the daytime. However, on my next investigation, I would find out first hand that this would all change, once the sun went down.

The cemetery is approximately three miles from the nearest small town, Palmer, Illinois. To get there, I had to first travel down a country road until I came to a green sign with an arrow directing me to Anderson Cemetery.

So much for Graveyard X being a secret location, I thought.

Turning left at the green sign, I then followed a narrow country road that winds for approximately another mile. Suddenly, there it was: the *elusive* Graveyard X. The mysterious cemetery, which I had heard and read so much about.

My plan for the day was to walk around, map out the location, snap a few photos, and use my digital voice recorder to see if I could record any EVPs. Nothing eventful happened that day, other than several flocks of wild geese and a bald eagle flying overhead. I took somewhere around fifty to sixty photos and only two of the photos had any type of anomalies in them. In both of these photos, there was a prism or rainbow color that appeared. The photos were taken at around noon, so the sun was basically straight overhead. It may have been nothing more than glare or a reflection, but I could not determine why the rainbow effect was only in these two photos.

It reminded me of the stories I had heard, in which people described seeing orange colored balls of light floating in the cemetery. One person told me how an acquaintance of theirs was in the cemetery after dark and witnessed these spheres of light coming out of the ground, filling a tree until it glowed.

As far as I know, I do not have any psychic abilities, but on this afternoon I had the strange feeling that this place was laughing at me. As if to say, "Sorry, nothing is going to happen to you this afternoon, but come back some night and we will show you what we can really do."

I decided to leave and to return the following Thursday night to see just what this place had to offer. I wanted to see if any of the tales and legends that I had heard about this place were true.

On Tuesday, March 6th, I made arrangements with a close friend, who was also interested in the paranormal, to help me with conducting the investigation. I decided that we would investigate the cemetery on Thursday evening and that we would arrive there at around 6:00 p.m. This would allow us to have enough daylight to set our equipment up and to get a feel of the surroundings before darkness set in. I knew that once the sun set; all of nature's natural sounds would probably change, becoming more mysterious and creepy.

Well, as my luck would have it, Thursday morning my friend informed me that he would not be able to assist with the investigation. I was so eager to investigate this location that I decided that I would proceed with it on my own. However, I did not relish the thought of being alone in a secluded rural cemetery that had the reputation of being extremely haunted and eerie after the sun went down.

Around 5:30 p.m., I loaded up my equipment and began the nearly ten-mile trek to the site. For some reason, I kept thinking about the old Don Knotts' movie, *The Ghost and Mr. Chicken,* in which a mild mannered, wannabe newspaper reporter spends the night alone in a haunted house. It was one of my all-time favorite movies, and I felt much like the Don Knotts' character, Luther Heggs.

This would be my first, but not last, experience in which I would conduct an investigation alone, in a cemetery. As I drove to the location, the sun was already starting to set, so when I arrived the shadows were already becoming more apparent. As I unpacked my equipment and began the short walk from the parking lot to the cemetery, I still did not have any real apprehension about the place. The birds were chirping. There was just a very peaceful and calm feel to it. This, however, was all about to change.

Upon arrival, I decided that the best use of my time, while it was still daylight, would be to walk around and listen to the natural sounds around me. I knew that once darkness set in, the simplest sounds of nature might seem to be eerie. I did not want to mistake a common noise, like the sound of a bird or the rustling of a tree branch, for something *paranormal*.

One book, featuring Graveyard X, described the most active area of the cemetery as a triangulated area, marked by a stone bench, a tall monument at the highest point of the cemetery, and an arched stone monument.

I had found this area of the cemetery on my first visit and decided that I would concentrate my investigation there for the evening. I noted a large tree, which was full of many chirping birds, near the stone bench. The birds were chirping so loudly that they were almost annoying. Everything else seemed calm and serene. However, after about thirty minutes, as soon as the sun set, the birds suddenly stopped chirping and things became quiet. It was as though they knew what was about to happen. It was as though someone simply turned off a switch.

The graveyard itself had an entirely different feel about it, as the day changed into evening. Only minutes earlier, it had such a peaceful feeling. But as the darkness grew, and I started walking around the graveyard, all was quiet. There were no birds chirping and no wind. It was so still that you could describe the feeling as *the calm before the storm.*

About forty-five minutes into the darkness of the evening, I was walking around the perimeter of the triangulated area, in the oldest part of the graveyard. I figured that from what I had read, if anything was going to happen, it would take place in this part of the cemetery. I stopped for a moment and began scanning the area with my eyes. To my left, I heard and felt someone or something walk past me, something that I could not see. I can best describe the feeling as like when someone passes you on the sidewalk. I could hear the movement and feel the breeze caused by whatever it was as, but it was invisible. Although I knew there was not anyone behind me, I felt that someone was there. I acted like I had not heard anything, slowly turned around, and took a photo with my digital camera. I reviewed the photo, but nothing was in the picture.

Then, once again, I heard the movement. This time, it sounded like it was to my left and a little in front of me. Then it passed by again, just like it did the first time and it seemed to go behind me again. I turned and attempted to take another photo, but my digital display flashed *low battery,* so the flash would not work. I was puzzled by the low battery message, as I had just put in new batteries before I had arrived. I then turned back and froze. I listened for more movement, and sure enough, there it was. It sounded again like it was directly behind me. Then, just as suddenly, there was complete silence, and when I say complete silence, I mean it was an eerie silence. Not even a branch or a breeze stirred. It was, as they say, *dead silent.*

I stood very still, listening and facing to the east, when all of a sudden someone, or something, punched me in the middle of my back. The punch was hard enough that it caused me to lose my balance and stumbled forward. I was wearing a t-shirt, sweatshirt, and my winter jacket that was unzipped about half way. My jacket was puffed out in the back but when I was punched, I was hit hard enough that my jacket was pressed against my back. An instant chill went down my spine. Immediately after I was punched, it felt as though something passed by me again to my left. I attempted to take a photo, but again, my flash failed.

Shocked, I stood very still and listened for any more sounds. After about five minutes, I was able to take several photos as my flash was working again, although my camera indicated that the batteries were still low. I walked around in the immediate area for another five to ten minutes. Then, something came over me. I am not sure if you can call it an intuition or a sixth sense, but I began to get a creepy, empty feeling.

Hey, you are all alone in a remote and secluded cemetery, I thought.

I was just punched in the back by something that I could not see and I had no idea what its intensions were. *Was it playing with me, or was it trying to tell me that it was time for me to move along?*

I decided to pack it up and call it a night.

Upon returning home, I reviewed my photos and there were ten or so photos that had very large orbs in them. Most of the orbs seemed to be located around a tree, in the center of the triangulated area, described earlier in this chapter, but there was nothing, which I could call actual evidence.

When I reviewed my digital recorder, there was one possible EVP, which sounded like it was either singing or chanting, but it was so faint that I could not use it as evidence.

Needless to say, I was pleasantly surprised by my first nighttime visit to Graveyard X. The strange feeling I had that Sunday before that I would experience something at night came to pass. When I first arrived, even after the sun had set for the evening, the place still seemed pleasant. I remember thinking that maybe all of the rumors were just that and I was going to be disappointed.

Even though I had heard the rumors and read the stories about Graveyard X involving all the unexplained lights and cold spots, I still arrived without any preconceived notions that can make the imagination run wild. But with whatever it was that had punched me in the back, I quickly became a believer that the stories and rumors about this place were true. What further confirmed my belief were the sounds of unexplained movement that I had heard. My experience was now firsthand, so I no longer had to rely on the stories of others to decide if this place really did come to life after the sun went down.

The vibe of a location can seem to change once darkness sets in, but for the first forty-five minutes after sunset, I noticed no real change in the feel of Anderson Cemetery. Soon after I began hearing movement, however, there was a whole different feel about the place. It felt like someone else was there with me.

I returned to the location the following Sunday during the mid-afternoon hours, and again, it had the original pleasant feeling about it that I had felt that first Sunday afternoon. There were no signs of any activity at all. I took photos and recorded audio for about an hour while there, but nothing out of the ordinary was present.

My next plan to try and obtain evidence was to leave a digital voice recorder overnight to see if I could capture any EVP evidence without being present.

A week later, on a Saturday night, I returned to the cemetery at approximately 8:30 p.m. Darkness had already set in when I pulled in to the cemetery. Upon getting out of my vehicle, I visually surveyed the graveyard for any signs of the unexplained lights that I had read about, and to check for any type of movement, but none was noted. I stood near my vehicle for a short time, listening for any sounds of the movement that I had heard on my first evening visit, and again nothing, not even a bird chirping. It was kind of an eerie quiet, which would soon change. As I walked through the opening in the fence, and took my first step into the graveyard, I felt like I was being watched from the north. I walked calmly and steadily through the cemetery, into the triangulated area where the stone bench was.

I placed a digital recorder near the base of the bench, took a look around, and there still was not any sign of activity. I turned around and began the walk back to my vehicle, all the while feeling like someone or something was watching me. I was almost to the fence and out of the cemetery, when I could not remember if I had turned on the recorder. So, I made an about face and again made the walk back to the bench, all the time feeling as though I was being watched from the same direction as before. I checked the recorder and, sure enough, I had turned it on, so I made a final walk back to my vehicle. No sooner had I stepped beyond the fence, and into the parking lot, the feeling of being watched dissipated. What was unusual about this was that even though I was now in the parking area, which was equally dark and secluded, I felt safe and secure. Stepping into the cemetery had caused a whole different *vibe*.

Bench with recorder at Anderson Cemetery (Graveyard X)

Before getting into my vehicle, I snapped a quick picture of the northern part of the cemetery with my digital camera and, immediately, I could see a large bright orb in the picture.

Once I returned home, I reviewed the photo on my computer and zoomed in on the orb. There appeared to be a face of a man in the orb. Some investigators believe that faces of spirits appear in orbs on occasion. I am not convinced yet that the faces are anything more than just mere anomalies and matrixing of distorted photos of moisture and dust. There is a psychological phenomenon known as *pareidolia*, which involves vague and random stimulus that is perceived as a form of *apophenia*, the experience of seeing patterns or connections in random or meaningless data. Common examples of this include seeing images of animals or faces in clouds, or the famous *man in the moon.*

In the photo that I took at Anderson Cemetery, the face looked very clear, but I still hesitate to call it evidence of spirit activity. Being punched in the back by something unseen, however, lends a little more credence to the possibility of the face in the photograph being paranormal.

It should be noted that, in the last eight years I have only taken three other photos that have orbs or mist in them, which I can classify as *unexplainable.* One was taken in the daytime around 7:00 p.m. at Williamsburg Hill in Illinois. It was a large, round orb and seemed to be giving off its own light. The photo was taken without the use of a flash. The second unusual orb was also taken at Williamsburg Hill and is not really an orb, but a red mist. The unusual thing about this image is that the mist has a leaf from a tree partially in front of it, which means that the mist is not dust or some other foreign particle in front of the camera lens, because it has a solid object between the lens and the mist. The third orb, or unexplained photo, was taken at the Villisca Axe Murder House in Iowa. I will discuss this topic in depth later in this book.

The next morning I returned to Anderson Cemetery and retrieved the digital recorder which I had placed under the bench. When I reviewed the audio, there were no unusual sounds that could be considered to be indicative of paranormal activity.

That October, I returned to Graveyard X with a local central Illinois paranormal investigator, named Ed Osborne. Ed is very knowledgeable in the paranormal field. He is a good friend and someone who I have a great deal of respect for. It was the Wednesday night before Halloween, and it was a cool, but not overly cold, evening. Ed and I had been in the cemetery for close to an hour, taking pictures, checking for electro-magnetic field (EMF) readings and monitoring temperature levels. Nothing seemed to be out of the ordinary. It was about 7:30 p.m. and the curfew for the cemetery was 8:00 p.m. The county sheriff's department patrols the cemetery, so we decided to make a final pass through, and then leave by curfew. Before leaving, I scanned the cemetery for a temperature reading, using my laser-pointed digital thermometer. The average temperature that night was forty-four degrees Fahrenheit. Everywhere I scanned, the temperature read forty-four degrees, that was, until I passed by the cement bench.

As I passed by the bench, the temperature began to drop. First the temperature dropped below forty degrees Fahrenheit, then it dropped below thirty degrees, then it dropped below twenty degrees. The temperature continued to drop steadily until it finally reached a low of minus sixteen degrees below zero. Ed and I could not believe what we were seeing, so to make sure that there was not some type of malfunction, or that the thermometer was set to Fahrenheit and not Celsius, I shut it off, then turned it back on. It was definitely set to Fahrenheit. I started scanning the cemetery again. The average temperature was still forty-four degrees everywhere, everywhere that is, except the spot next to the concrete bench. Again, the temperature dropped below zero. My hands were freezing and my nose felt like it was becoming frostbitten. Every time I would move the thermometer more than four or five inches, the temperature would go back up to forty-four degrees, but as soon as I would move it back to the cold spot, it would go below zero. At approximately 8:15 p.m., Ed and I decided we had better leave since it was after curfew and we were trying to be respectful of the cemetery rules. When we left the cemetery, the temperature was still minus eleven degrees Fahrenheit. When we got in my vehicle, I turned the heat on to warm up my fingers and nose, because they felt frozen. So, at one point, we had experienced a sixty degree temperature drop, without explanation. I have been back to the cemetery many times since, but have never recorded such an extreme temperature fluctuation again.

Investigation Summary

Anderson Cemetery has more than lived up to its reputation, and I have no doubt that something supernatural is taking place there. In the oldest part of the cemetery, you get the feeling that you are constantly being watched or followed; a feeling that something is not right, but you just cannot figure out what it is.

I cannot place my finger on it, but each time that I return to Graveyard X things seem different than the last time I was there. I was starting to think I was imagining this, until one day Ed Osborne raised the issue with me. He said, "Larry, have you ever noticed, or is it just me, but every time I come back to this place, something seems different? I don't know what it is, but things seem to change." I explained to Ed that this was exactly the way I felt each time that I returned to Anderson. I have been there many times now and several times it seemed like tombstones had moved, or there were gravestones that had not been there before, even though I had walked past that particular spot many times. I am not talking about new stones, but very old stones. It seemed like the physical environment of the cemetery had been altered.

Graveyard X is a very odd, but a beautiful place. If you have never investigated it, you definitely need to put it on your list of places to visit. One thing to remember is that there is something there that lurks in the shadows of the cemetery. It seems to come alive and interact with visitors at night. I experienced this firsthand when I was punched in the back by some unseen force, a force which seemed to have intelligence about it and one that wanted to let me know that it was there.

If you decide to make the trip to rural Christian County to see for yourself, make sure that you get there early, because after the sun sets, Graveyard X seems to come *alive*.

Ridge Cemetery at Williamsburg Hill

CHAPTER TWO

Williamsburg Hill (Ridge Cemetery)

Tower Hill, Illinois

"Fear...it's the oldest and strongest emotion known to mankind...and our greatest fear is the fear of the unknown." H.P. Lovecraft

The above statement, by science fiction writer H. P. Lovecraft, is so true when it comes to investigating the supernatural. One of the great unknowns has always been the age-old question. *Is there life after death?* Storytelling is one of ways that helps us to understand the realm of the unknown. It also helps to explain what may await us, once we make the transition from life as we know, to life after death. I for one have always been too curious to simply leave it up to stories and legends, preferring to find the answer to this question myself. I do this, by exploring places that most people seem to shy away from, due to their own fears.

Ridge Cemetery at Williamsburg Hill is one of those places.

As motorists speed by on their daily routine, driving down Route 16 in rural Shelby County, they see what some would call a pleasant existence. If you asked them what they saw on their travels, you will often be met with laughter. The landscape is a maze of cornfields, small towns, and modest farms. The world outside their windows is perceived as being peaceful and normal. Their daily trips are considered mundane and uneventful. When you pass through the small town of Tower Hill, nothing looks out of the ordinary. But if by chance you exit and turn off the main road traveling through the countryside, following one of the many narrow roads in the region, one road will lead you to a whole different world. A world that goes unnoticed, when you are traveling at sixty miles per hour.

County Road 1100E leads you to one of the strangest mystery spots in central Illinois, *Ridge Cemetery* at *Williamsburg Hill.* Located in the south central part of Illinois, near the small community of Tower Hill, the cemetery is not hard to find, as it sits on a ridge at the very top of the hill. The hill itself is unusual, in that it stands 810 feet high, making it the highest elevation in downstate Illinois.

At one time, there was a thriving village at the top of the hill. *Williamsburg*, also called *Cold Spring* sat on a ridge near the top of the hill. Dr. Thomas Williams and William Horsman founded the village of Cold Spring in 1839. Several of the Horsman family members are buried on Williamsburg Hill both in Ridge Cemetery and on private property located on a farm just on the other side of the hill. For some forty years, Cold Spring survived as a thriving village. It housed a blacksmith shop, doctor's office, two churches, and a saloon.

One of the primary reasons that the community thrived was due to a stagecoach route that ran from Shelbyville to Vandalia and passed through the village. This brought prosperity to the small community for years, but in 1880, when the Beardstown Shawnee Town and Southeastern Railroad were built, the owners of the railroad decided to bypass the hill. This caused the stagecoach line to die out and, in turn, so did the village of Cold Spring.

Today, very little is left of what once was the thriving village of Cold Spring. The remnants of the village are covered by trees and underbrush, hidden from view. A few families still live nearby, secluded from the hustle and bustle of the big city life, but, *do they live alone?*

Hidden amongst the trees and underbrush is a very odd place, *Ridge Cemetery.* Although located off the main road, the graveyard is easily found simply by looking for a tall microwave tower that can be seen from miles away. Once you arrive at the tower, turn immediately left and follow the road about a quarter mile. You may notice that, while traveling down the narrow gravel road, you feel like you are driving through a tunnel. This is due to the close proximity of the trees in this highly forested area. You may also get the feeling that you are in the middle of nowhere, and in some respects you are right. But what you do not expect is to see what awaits you at the end of the road, a place that I have found to be shrouded in mystery.

When I first heard the stories surrounding Ridge Cemetery, what I expected to find was a desecrated graveyard overgrown by weeds with broken and damaged tombstones. However, to my pleasant surprise, the cemetery is very well maintained. But well maintained or not, I classify the location as one of the strangest and oddest places I have visited in the years that I have been a paranormal investigator.

As an investigator, I have been to some pretty secluded locations and cemeteries. Many of these I have investigated alone. But what separates Ridge Cemetery from other places is the feeling that it sometimes gives off. Many times, I have had the feeling that I have taken a step back in time. It is like I have discovered a lost world. A lost world that is alive with some type of spiritual or cosmic intelligence.

If you venture down the road at night, you will find a very dark and frightful place. The shadows seem to move and come alive. That, along with the feeling of being watched, has been enough to keep most away after the sun sets.

I have been in the cemetery many times, during both the day and at night. I never know what to expect when I get there. On some occasions, I have arrived at the cemetery and everything felt completely normal, as it should be. Other times, I have had the feeling that there is a presence. A presence which does not want me to be there.

During research and investigations for a documentary I am working on called, *Strange Williamsburg Hill*, I have talked with many people. One was a lady named Cathy, who has been a lifelong resident of central Illinois and who has been coming to Williamsburg Hill for over thirty years. She told me that she has experienced many odd things both in Ridge Cemetery and in the timber that surrounds the graveyard. I will explain some of her incredible encounters more in depth later in this chapter, but one of her stories in particular caught my attention. It concerned a day in which Cathy, her daughter, and six year old granddaughter decided to take a trip to Williamsburg Hill, to see if they could further experience some of the oddities that the hill had to offer. She told me about how they had even packed a picnic lunch so that they could spend the entire day there.

According to Cathy, they had only been there for ten minutes or so. They had laid out a blanket, and had started to unpack their lunch. Suddenly, Cathy and her daughter had an overwhelming feeling come over them. Eerily, both Cathy and her daughter had the strange feeling, at the same time! It was a feeling that they were not welcome and should leave, which is exactly what they did. They packed their food and blanket, loaded everything up in the car, and left. For one person to have a bad feeling about a location is one thing, but when two people simultaneously get the same feeling, it makes you wonder if some sort of collective consciousness or possibly *some other force* was warning them that they were not welcome there. Cathy told me about other times she has been there. During those times, she had a feeling of peace and that she was welcome. But not on the particular day that she described.

Williamsburg Hill is a location that I have investigated more than any other. The reason for this is that I am working on a documentary about the hill and what takes place there. The film is called *Strange Williamsburg Hill*. No date has been set yet for the release of the film. Always a perfectionist, I have filmed and re-filmed it using high definition video and I am in the process of re-editing the film once again. I hope to have it completed sometime in the spring of 2015. Over the course of filming, I have investigated the location some one hundred times, so I have a pretty good sampling of how it feels at various times and conditions.

I have read stories about unmarked graves being found in the timber that surrounds the graveyard, but as of yet, I have not seen any evidence of such things myself. Other tales, that I have heard, such as strange unexplained noises, cattle mutilations, big cat sightings, unidentified lights, apparitions of an old man and stories of a woman who dresses in black, all sound like something out of a Stephen King novel. But I have talked to eyewitnesses, who confirm all of these stories.

The old woman who dresses in black has been seen inside the cemetery and a well-dressed elderly man has been seen on the road leading up to the graveyard. Both appear out of nowhere and seemingly vanish into thin air. During the course of interviews for the documentary, I have talked to two people who have firsthand knowledge of the lady in black. Another person that I interviewed claims to have actually held a conversation with the old man, who then practically vanished before her eyes.

Good friend and fellow paranormal investigator Ed Osborne, is one who has witnessed the lady in black. I had mentioned to Ed some of what I had experienced out at Williamsburg Hill, and it was enough to intrigue him to take a road trip there, with one of his friends, to look around for himself.

Ed told me that when they arrived he figured that his friend and he were the only ones there, since no other cars were in the parking area near the entrance of the cemetery. There is only one way in and out of the cemetery *or at least there is only one way in and out of the cemetery if you are a living breathing soul.* However, when they walked into the cemetery, they noticed that in the far corner, what appeared to be elderly women standing near a graveside.

The odd thing that Ed noticed was that she did not even look at them or seem to be concerned that two strange men had suddenly arrived in such a secluded setting. An elderly woman would, or so they figured would at least want to keep an eye on them, especially not knowing what their intensions might be.

Ed and his friend began looking at some of the older headstones in the cemetery. It had only been a moment, when Ed glanced back to where the woman had been standing. She was gone. When Ed explained the story to me, he said, *"Larry, she just seemed to vanish, because she would have had to pass by us in order to get to the cemetery gate."* Ed was right, because the cemetery is surrounded by thick timber and brush. With the way she was dressed in a long skirt, she could not have possibly made it through the timber, without becoming tangled in the underbrush. Plus she appeared to be elderly and would not have been able to maneuver through the treacherous landscape by herself. Ed stated that he looked all around and the woman was not in the cemetery, nor was she on the gravel road that leads to the cemetery. No other cars were in the parking lot and no other cars came to pick her up. He was completely baffled as to where she could have gone.

As Ed was telling me the story, it reminded me of another story that one of the locals told me. This local's six year old granddaughter had also encountered an old woman in a black dress. So, as Ed was telling me the story, I asked him to describe the old woman to me. Ed described her as elderly, wearing her hair pulled back in a bun, wearing a black dress or skirt that appeared to be from an earlier time period. When he told me the story and described the old woman, he had no knowledge of the girl's encounter.

That encounter, was told to me firsthand by one of the locals who has been coming to Williamsburg Hill and Ridge

Cemetery with some regularity for years. The story took place around 2006. On the particular day that the encounter took place; she had brought her granddaughter along with her to the cemetery. Her plan was to walk around and check out some of the older stones in the graveyard. When they arrived and entered the cemetery, her granddaughter wandered off to play in the far right-hand corner of the cemetery. The grandmother told me that it was only a few minutes before her granddaughter came running back, terrified. She asked the little girl what was wrong, and her granddaughter excitedly explained, that an old woman had asked her if she *"wanted to play with the children."* When her grandmother asked, *"What woman are you talking about?"* The girl replied, *"The woman over there in the black dress."* The grandmother told me that as she looked to where her granddaughter pointed, there was no one there. Still frightened, her granddaughter, said, *"When I asked the woman where the children are"*, she told me, *"They are under the ground!"* As soon as her granddaughter said this, a chill ran down her spine. The grandmother went to the far corner of the cemetery to confront the old woman, but when she did, there was no one there and no one else was in the cemetery. She told me that in order to leave the cemetery the old woman would have had to pass by them, but she had not, and if she was an elderly woman, there is no way she could

51

have made it through the rough terrain of the hilly timber that surrounds the graveyard. It was as though she had simply vanished.

The spot where the little girl encountered the old woman and the spot where Ed Osborne saw the old woman in the black dress were in the exact same locations. I truly believe that they both witnessed the same specter.

Cathy, who is the same Cathy who had abandoned her picnic at the cemetery during an earlier visit, told me what is perhaps the strangest story that I have ever heard concerning Ridge Cemetery. In April 2007, her husband Jerry and she had traveled out to Williamsburg Hill to do some mushroom hunting.

She told me when they had arrived they backed up their pickup truck to the fence near the gate of the cemetery. They then headed into the timber off to the right. Cathy said her husband had just made his way down a small embankment when he asked if Cathy would get him a drink out of the cab of the truck before she came all the way down the embankment. Cathy made an about face and headed back to the truck to retrieve a drink for her husband. To her surprise, as she made it back to the parking lot, there was an elderly man standing in front of their truck. This startled her, as he had not been there before and because she immediately noticed the odd way in which the man was dressed. It was odd because he was wearing brown pleated pants and a silk button up shirt. Cathy described his hair as a beautiful grayish silver color and was neatly combed back. He was wearing shinny wing tipped shoes that looked freshly polished. She told me that the man was dressed like her grandfather would have dressed and guessed that he looked like someone out of the 1920s. It also puzzled her that the man's shoes were so clean, since it was during the springtime, and it had rained recently.

She said that he should have at least had dirt or traces of mud on his shoes, but they were spotless. Plus, he had to have walked there since there were no other vehicles around. Cathy was also puzzled by the fact that she and her husband had just arrived and had not seen the man on the road. As far as they knew, there had not been anyone in the cemetery either.

When she approached the truck she said hello to the man, but he did not reply. She became alarmed when he did not respond to her greeting, so she once again said hello. This time the man responded, but Cathy could not understand what he was saying. She then asked him if she could help him with something. The man mumbled in response. *"Excuse me?"* Cathy replied. Finally, in a voice she could understand, the man asked, *"Can you tell me where the bars are?"* When Cathy was telling me the story, she told me that the odd thing was that when the man spoke, his lips and mouth did not move. But she had heard him speak and understood exactly what he had said.

Cathy wasn't sure what to tell the old-timer, as she wasn't sure where the closest bar was. They had some beer in a cooler in their truck, so Cathy offered him a beer. Once again, the man responded with the same question, *"Can you tell me where the bars are?"* She again explained that the man's lips and mouth did not move when he spoke. Cathy told the man to wait and that she would go ask her husband the location of the closest bar. Cathy hurried down the small embankment and was greeted by Jerry, who said, *"What does this guy want?"* Jerry had heard the man, but had not given it a second thought. Cathy explained what he said and also told her husband that he needed to see this guy, because she knew that something was not right. Both Jerry and Cathy made their way back up the embankment, but to their surprise, the man was gone. He was nowhere in sight, Cathy told me. He was not on the road, because, from that vantage point, the road is visible for a good eighth of a mile. They checked the cemetery and he was not there either. They knew that he was not dressed to walk through the dense and muddy timber that surrounds that cemetery, plus they would have definitely heard anyone moving around in the thick underbrush.

About a month after the incident with the old man, Cathy was talking with a visitor that she met at Williamsburg Hill. She said he had maps of the area and seemed to know a little about it. Cathy had not mentioned her encounter with the old man when this fellow started telling her about a story he had heard concerning a traveling salesman who lived in the area. It seems that the salesman had an argument with his wife, so he headed out to have a drink at a local bar. Unfortunately, he didn't make it to the bar; because he was killed in an automobile accident on the way home. Could this be the man that Cathy encountered? Is the salesman now some lost soul trying to find his way to his final destination?

I met a man named Jason during one of my many days of filming at the cemetery. He is one of the caretakers of the cemetery, and told me of another strange encounter that happened to him. According to Jason, his mother and he had just finished mowing the cemetery and he had parked his John Deere lawn tractor atop the hill near a very large oak tree in the center of the graveyard.

Oak tree at Ridge Cemetery at Williamsburg Hill

He explained that after they finished mowing, his mother decided to take a walk down the backside of the cemetery, to look at some of the older tombstones. Jason parked the tractor by the oak tree, at the top of the hill, and made sure that it was in gear, so that it would not roll away. He then joined his mother and they took a short walk down the hillside at the rear of the cemetery. A few minutes later, they returned to the top of the hill and, to Jason's surprise, the tractor was gone. He reassured me that the tractor did not simply roll away, as he had left it in gear. He told me that if someone had started up the lawn tractor they would have definitely heard it, as it was a very quiet Sunday afternoon. They also would have heard the tractor if someone had tried to push it down the hill. The tractor was simply nowhere in sight.

Finally, after searching for several minutes, Jason and his mother found the tractor in the timber approximately seventy-five yards away from where he had originally parked it. He said that the tractor was sitting in the timber and turned facing the direction that it would have come from. The tractor was still in gear, unscratched and unscathed as if it had been parked there on purpose. Now, this is where the story gets even stranger. In order for the tractor to get from where it had originally been, to where it was now sitting, it would have had to have popped out of gear, rolled down the steep hill, while avoiding several hundred tombstones, large trees, and branches. Then it would have to have put itself back in gear, and end up where it was found without causing any damage. Jason said that it was almost like something had lifted up the tractor and placed it in the timber. To this day, he still cannot figure out what happened.

Jason also told me about other caretakers who simply refuse to work out at Williamsburg Hill due to feelings of being watched and followed. He said that these same caretakers are used to working in other cemeteries, but there is something about Ridge Cemetery that simply frightens them.

I also interviewed several local residents who lived out at Williamsburg Hill, but did not wish to appear on camera for my documentary. One of the residents was the great granddaughter of one of the founding fathers of Williamsburg / Cold Spring, William Horsman. She was not a big believer in the supernatural and she resented the stigma that was tied to the hill. But, after expressing this to me, she told me a story that she remembered as a child.

Her father, who lives in a nursing home, was an eyewitness to the event. In 1969, a local farmer found several of his cattle dead along a creek bed. It was not apparent how the cattle had been killed, but their reproductive organs, eyes, and tongues had been cleanly cut out and removed. Almost as if they had been surgically removed. The cattle were found near a muddy creek bed and there were no footprints or signs that an animal was responsible for the deaths of the cattle. All of the blood had been drained from the carcasses. This type of activity is found in most classic cases of cattle mutilations.

I also talked to another local who told me of a time that he was just south of Ridge Cemetery and was hunting in the wooded area at night. All of a sudden, a large orange ball of light hovered in the sky just above the tree lines. "It wasn't an airplane or a helicopter," he told me. "It was just a light and it didn't make any sound." He said that the light simply disappeared. He told me how only a few minutes later he witnessed four or five fighter jets as they skimmed the treetops. They acted as though they were looking for something. His conclusion was that they were looking for the strange light.

On July 10th, 2010, I was investigating Ridge Cemetery with a couple of local paranormal investigators when one of the investigators, Chris Mason, looked up and saw a large orange orb or ball of light traveling above the tree line. The object appeared to be on a course to land nearby.

Example of Orb at Williamsburg Hill taken during daylight without flash

On the same night, we ran into a teenager who was accompanied by his aunt. They had traveled to Ridge Cemetery from Mattoon, Illinois because they were curious about all the stories that they had heard about the mysterious location and were hoping to experience some of the activity for themselves. They were excited because they too had seen a strange light that was similar to what Chris had seen. They described it as acting like a fish swimming around in the sky. Their description of the object was identical to what Chris saw; with the exception that the light they saw was a neon green color. The boy described the light as floating around erratically. He further described it as acting like a fish swimming in water.

On another occasion, an eyewitness told me about the night he and a friend had driven out to the cemetery to hang out. It was dusk so it was fairly dark. He admitted that they had consumed a couple of beers, but was adamant that they were not intoxicated. They had been at the cemetery a good forty-five minutes or so and were sitting on the tailgate of his truck. Suddenly, something big began rustling in the large oak tree in the graveyard. The tree was approximately forty yards in front of them. Whatever was moving around was big and it was in the tree. All of a sudden, a very dark, what they believed to be black shadow leapt out of the tree and landed approximately twenty-five yards away. The eyewitness told me that whatever this was then leapt again and it landed in the timber. This meant that it had leaped another twenty-five yards. After this thing landed in the timber, it was completely silent. No movement or sounds that you would have expected from something as large as they had seen. His first thought was that a black panther had jumped out of the tree. He told me that he had heard stories of big cats and panthers being seen out on the hill. I asked if they saw it any more that night, and he said, *"Heck no! That was all we needed to see, and we left right after that happened!"*

Last summer, Chris Mason and I talked to a resident who lives near Ridge Cemetery. He told us of a time a few years ago that he heard and saw a mountain lion walking in the timber behind his home. The man had been burning a pile of old papers. The pile was fairly large, so when he lit it, the fire quickly crackled to life. In a sort of response to the sound of the fire, he heard the roar of a big cat. He believes that the fire startled the cat. The witness told us that just behind his house there is a large drop off of about forty feet, and at the bottom of the drop-off is an old wrecked station wagon, that he believes the cat was using as a den.

During the filming of my documentary, I experienced several things that I would have to classify as unexplained or quite possibly paranormal.

In June 2008, I journeyed out to Williamsburg Hill at night for the first time in order to shoot a couple of hours of footage at the cemetery. It was already dark when I arrived, so the first thing that I did was to carry my equipment up the hill to the top of the cemetery. I laid it near the big oak tree in the center of the cemetery. I then began to set up my tripod and video camera in order to film a variety of shots from different angles for the film. I had only been filming for about thirty minutes, when I decided that I would move the video camera down near the gate, at the entrance to the cemetery and film up the road. In hope that I would be lucky enough to catch the apparition of the old man.

As I was removing my camera from the tripod and getting ready to move, I heard whistling coming from the woods. The whistling was off to my right and sounded like it was about forty yards away. The strange thing was that it was coming from the woods. It startled me because it was nighttime and I was in a secluded rural cemetery that is surrounded by dense forest and what would be considered treacherous landscape even in the daytime. Add to that the fact that it did not sound like it was coming from a bird, but a person. It was the type of whistling that a person makes when they take two fingers and place them under their tongue and make a *come here* sound. I stopped in my tracks. Everything was still; the birds had quieted and roosted for the night, long before I had arrived. I had the feeling that I was no longer alone and that someone in the forest was watching me. I talked myself into believing that I had not heard the whistling, so I grabbed my tripod and camera and headed toward the gate. I was about half way between the large oak tree and the fence when I heard the whistling a second time. This time, I had definitely heard it and it was not a bird or an animal. It was if a person was whistling for me to *come here, to come in to the woods*. What confused me was that the whistling was now coming from an area about twenty yards closer to me, than the first time. So either it was moving closer or there were two things whistling in the woods for me to *come here*.

In my mind, I knew that no one could be in the timber at night, without a light, because it was pitch black in the cemetery. Even I was using a flashlight. No one could possibly navigate in the woods due to the darkness and terrain that they would have to contend with. If someone was in there, I should have heard them moving around. Plus, if they did have a light, I would have seen it. I tried to remain calm and acted as though I had not heard the whistling, because if someone was in the woods, I did not want them to know that I knew it.

I continued walking down the hill and went through the gate. I set my tripod up in a spot about ten feet from the gate and another ten feet from the timber off to the right, as I faced down the road.

As I was setting up my video camera on the tripod, the whistling began again about ten feet away from the edge of the cemetery. This time there was no doubt, what I heard. It sounded exactly like a person making that *come here* type of whistling, and this time it was just a few feet away from me. There is no way whatever was whistling for me to come into the woods could have moved around without me hearing them. I have been in the timber during the daytime and with every step that you take, you make loud crunching and snapping sounds due to the dense underbrush. To do this silently, in complete darkness, would have been impossible, even with night vision goggles.

I tried to remain calm and continued acting as if nothing was wrong. I decided that I had better head back up to the large oak tree, as I had left the rest of my equipment under there. If someone or something was lurking in the dense underbrush and I needed to make a fast getaway, I sure did not want to leave expensive equipment behind.

As I headed back up the hill toward the tree, the whistling happened at least ten more times before I made it there. Once again, the whistling moved with me, following me up the hill, always in the timber. I had the feeling that whatever was there was inviting me or daring me to come into the woods.

When I made it back to my equipment, I opened one of the bags and took out my laser guided digital thermometer. I turned it on and hit the light to illuminate the screen so that I could see the temperature reading in the darkness of the night. The average temperature was about eighty degrees, but when I pointed the thermometer in the direction that I last heard the whistling, the temperature began to drop. It slowly dropped into the upper seventies, then the lower seventies, then into the sixties, and then, all of a sudden, it hit sixty-six point six degrees Fahrenheit and locked on it. Now, these thermometers are made to scan and I have never before or since had this or any thermometer of this type lock on a number. The thermometer stayed on sixty-six point six degrees Fahrenheit for almost thirty seconds. For the uninitiated, *666* has long been thought to be the *number of the beast*, a satanic number. That was all I needed to see.

I packed up my equipment and headed back down the hill to check on my video camera. As I made the long walk down the hill, again the whistling followed me. I was now sweating and very concerned. After all, I was out in the middle of nowhere, down a dark narrow gravel road, and surrounded by dense timber. One thing that kept going through my mind was that there are Native Americans buried in unmarked graves in the cemetery. I had read and seen many documentaries about Native Americans. I recalled how they would communicate with each other by whistling and how they were able to move around in the forest in a stealthy manner. If it was a spirit, or something else supernatural, it quite possibly could move around without disturbing anything.

Finally, I made it back to my video camera near the entrance of the cemetery. I noticed that the LCD screen was black, but it was still recording. The only time the LCD screen on my camera should go black is if the batteries run down. I had over ninety minutes of battery left, so this did not make sense to me. As I was checking out my camera, I heard the whistling again and it was nearby, probably less than ten feet away. That was the last straw, I felt like I had pushed my luck far enough for one night. It was so dark out there that I could barely see my hand in front of my face. I grabbed my equipment and did not even bother taking the camera off of the tripod. I just laid everything on the front seat of my vehicle. I placed the key of my SUV in the ignition and remember saying to myself, "Please start!" A sigh of relief fell over me with the roar of the engine. I went down the road, leaving this dark and frightful place in my dust, knowing that I would be back another day.

As I continued heading home, I was fifteen or so miles away from Williamsburg Hill near the small town of Pana, when I heard a noise on the front seat of my vehicle. It was the LCD screen on my video camera. It was lit up and was working fine. When I left the cemetery, I had forgotten to turn off the camera, but now it seemed to be working again.

Once I arrived home, I reviewed the video footage. I was using an infrared light on the camera with *nightshot* and had hoped that I may have captured some evidence, but all of the footage of the gravel road turned out black. There was audio, but even though the whistling came as close as within ten feet of the camera, it was not recorded on the camera's microphone, which was very disappointing to me, to say the least.

That was a night that I would remember for a long time.

I have been investigating Williamsburg Hill since 2006 and, as mentioned previously, I have been there many times both in the day and at night. There have been so many unusual experiences in and around the cemetery that it would take an entire book to discuss. So, I will summarize some of the odd things that have happened while investigating the hill. Then, I will relate one more story at the end of this chapter in which I conducted an investigation with my friend and central Illinois paranormal investigator Chris Mason, as well as well-respected and well-known Bigfoot researcher, Stan Courtney.

One of the most interesting things that happened to me was an EVP that I recorded while investigating with another good friend and paranormal investigator Jamie Sullivan and his two sons, Danny and David. It was in April 2009. We did not have a lot of personal experiences that night, other than we kept hearing what sounded like wood knocking and at other times what sounded like someone was taking a heavy log and dropping it on the ground. We shined our flashlights into the timber and, of course, nothing was there. I was carrying a digital audio recorder as we walked through the cemetery. When I returned home and reviewed my audio, I found that I had recorded an EVP. The EVP is a whisper that says, "Wellman!"

One month later, in May 2009, Jamie and I traveled to Villisca, Iowa to investigate the famous Villisca Axe Murder House. It was Jamie's second investigation of the house and my third. When we returned from the Villisca trip, to my shock and amazement, I had again recorded a voice that whispered, "Wellman!" The intriguing thing is that I recorded the voice using a different digital recorder and recorded the same voice over 300 miles from Williamsburg

Hill. The only common denominator was that both Jamie and I were on these two investigations. We conducted the investigation in the Villisca house alone. I am not sure how this is possible or why, but it happened. Is this proof that spirits can travel from one location to the next, or are they somehow always with us?

There is one particular area of Ridge Cemetery that seems to have more activity than the rest. This area is located in the far right-hand corner of the graveyard as you walk through the gate to cemetery.

Last summer, on July 10th, the same night Chris Mason saw the strange light in the sky over the timber; we had numerous other unexplained things happen. The investigators with me that night were Chris Mason and Janet Morris. Janet was fairly new to paranormal investigating, but, being *sensitive*, has been in touch with the paranormal and spiritual side for years. I will not classify anyone as being sensitive, unless they have, in some way, been able to prove to me, with some type of tangible or measurable evidence beyond any reasonable doubt, they are for real. In the short time that I have known Janet, she has proven to me her legitimacy time and again through her accurate statements, about things, of which she had no prior knowledge. On this particular night, Janet stopped me dead in my tracks with a vision that she had. Chris, Janet, and I were walking toward the far right-hand corner of the cemetery. This was Janet's first trip to the cemetery and she was not aware of, nor had prior knowledge of, any of the stories that I had been told about Williamsburg Hill.

As we were walking toward the far corner of the cemetery, Janet suddenly stopped and said that there was an old woman standing near us. I asked her if she could describe what the woman looked like. She described her as being elderly, with hair pulled back in a bun, and that she was wearing a long, dark navy blue dress. The description that Janet gave was basically identical to the elderly lady that Ed Osborne had seen months earlier. The description was also very similar to the description that the little girl had provided to her grandmother of the old woman who had asked her if she "would like to play with the children." The only difference between the various descriptions was that Janet's old woman was wearing a navy blue dress and the dress that the old woman was wearing that Ed and the little girl described was black. I wondered whether, in the sunlight, a navy blue dress may appear to be black. The area that Janet was standing in was the same location where both Ed and the little girl had seen the old woman. Janet's vision of the old woman seemed to confirm that the *woman in black* existed.

That night, we experienced several strange things that we could not explain. After Janet's vision, we continued walking to the far corner of the cemetery. Chris decided to sit on a wooden bench, next to the fence line, directly in front of the timber. Janet and I had two folding chairs. At night, Ridge Cemetery is very dark and the timber that surrounds the cemetery is even darker. We had been in this location for maybe ten minutes when Chris stood up and said, "It feels like there's something standing behind me!" He turned around and asked me to shine my flashlight behind him. I stood up and shined the light in the direction he had indicated, but there was nothing there.

A few minutes later, we heard movement in the timber. I again shined my flashlight in the timber and, again, nothing was there. When I shined a light into the timber, the movement stopped. I picked up a walnut that was lying on the ground and threw it into the timber to see if it would cause anything to move. Within a few seconds, the walnut came flying back and landed near Chris. Chris and I looked at each other. We threw a couple of more walnuts into the timber, but this time nothing happened. I then took a photo of the area behind the bench. When I reviewed it, there was a strange, bright vibrant red mist in the photo. It was not dust or moisture, because part of the mist was behind a tree branch, which would mean that the branch was partially between my camera lens and the strange red mist. I have taken pictures with mist in them before, but never any mist that was bright red. There was nothing red in the timber or around us that would have caused a red reflection. To this day, we cannot explain the mist.

Several other odd things happened that night, including hearing wood knocking and the growl of what sounded like a cougar or some type of big cat. In the same area of the timber, we heard a muffled scream that sounded like a woman. It was a very faint scream. Moments later, we heard what sounded like multiple people screaming. Again, it was very faint. The screaming continued for a while and we were finally able to figure out where it was coming from. The screaming seemed to be coming from under the ground in the timber and very close to where we were standing. Was the screaming that of lost or tortured souls, or possibly a residual of something horrible that may have occurred on this spot at some previous time? Or, as some believe, is there a portal or vortex in the forest that surrounds the cemetery, and does it allow us to interact with other planes of existence?

Another example of an Orb at Williamsburg Hill taken during daylight

Chris and I investigate together often. When we experience strange or unusual phenomena like those screams, we try to rationalize it and look for logical explanations for what we are experiencing. That night, we could not find a rational explanation for the screaming or the walnut being thrown back at us. We lit up the timber with our flashlights and there was no one there, at least not in a physical sense. As for the screaming, it was definitely coming from under the ground.

Later that same night, around 3:00 a.m., we started hearing what sounded like Native American drumming, followed by chanting. Again, the sounds were coming from the forest surrounding the cemetery. The drumming and chanting continued for quite a while and was still happening when we left the cemetery sometime after 3:30 a.m.

After the experiences that we had on July 10th, Chris mentioned some of the noises and strange knocking sounds that we had heard to Stan Courtney. Stan is well respected in the field of cryptozoology and is considered by many to be one of the foremost experts on Bigfoot. His research takes him all over the country. He has been to many remote wilderness areas, conducted hundreds of investigations, and has recorded the sounds of creatures of the night with the use of sophisticated equipment. On most occasions, Stan will take his faithful companion, a Karelian Bear Dog, Belle.

When Chris told Stan about the activity in the timber around Ridge Cemetery, Stan explained how wood knocking and the throwing of objects were classic signs of Bigfoot activity. Just a quick note regarding Bigfoot. Stan is under the belief that Sasquatch, or Bigfoot, is some type of primate or possibly an undiscovered hominid or primitive creature. Chris and I have had many discussions on this subject and we both have come to the consensus that these large, hairy creatures, which have been seen over the years, in all parts of the world, may be some type of being that, for whatever reason, is able to breach the laws of time and physics, allowing it to go back and forth between time plains or dimensions. One of the reasons that we have this belief is that there have been many eyewitness reports of people seeing a Bigfoot type creature, describing them as being seven to eight feet tall and weighing over seven hundred pounds. That same eyewitness will also claim that the creature walked behind a tree and then seemed to disappear. Well, for one thing, something that big cannot hide behind the average tree. Add to that the fact that many times tracks will be found along muddy areas, and the tracks seem to be heading in a certain direction, then all of a sudden the tracks stop, like the creature has simply vanished.

Our theory regarding Sasquatch, and possibly why they have been so elusive, is that these creatures are simply not hiding behind trees and vanishing into thin air. We believe that they are stepping through some type of inter-dimensional doorway or portal and can somehow breach the laws of physical nature, to which we as humans are bound. Native American legends, and even the Bible, speak of giants that roamed the earth.

After talking to Stan, Chris gave me a call. He said that Stan would like to do an investigation and set up some audio equipment in the timber around Ridge cemetery. I agreed and we scheduled one for the following Saturday night.

That night Chris, another local paranormal investigator named Dan Wahl, and I met Stan Courtney and his dog Belle out at Williamsburg Hill. It was not a particularly active night for the hill, but something in or around the cemetery seemed to affect Stan's dog. Stan has described Belle as being a somewhat curious and fearless dog. Belle has been trained to be quiet and not bark when accompanying Stan on his investigations. He also told us that Belle will run around an area and explore it herself.

As the night grew dark, we began our investigation. As a group, we headed toward the corner of the cemetery where the majority of the activity has taken place. This was the same corner that I had thrown the walnut into the woods, only to have it thrown back at us. It was the same place that we had heard the screams coming from under the ground. It was also the same place where Janet had her vision, as well as where both Ed Osborne and the little girl had seen the old woman dressed in black.

We headed toward the corner of the cemetery; the four of us spread out and were walking side by side. I noticed that Belle was walking behind me and was staying right on my heels. I took a few steps to my left to look at something. At which point Bell hurriedly got behind me again and stayed very close to my heels. She stayed behind us the entire time that we were in this corner of the cemetery. If she was behind one of us and we suddenly moved, she would hurriedly move to get behind us again. It seemed as if something was scaring Belle. Stan explained to us how this was very out of character for her and that Belle was used to being in areas where bears and big cats lived and she would roam about freely without any fear.

Whatever was in that far corner of the cemetery, Belle did not like at all. We moved away from there and followed the fence line around, back toward the entrance to the cemetery. As we got closer to the cemetery entrance, we heard movement in the timber near where the gate is located. Belle headed toward the sound and seemed to either sense or see something in the timber. She lowered her head and body as if she was on point. She started growling and barked, but Stan instructed her to be quiet. He has explained to us several times since this investigation just how out of character Belle's actions were.

At one point, later in the night, we took a walk down the dark and secluded road that lead up to the cemetery. We used flashlights to find our way up the road, and as we were walking back, the beam from Chris' flashlight lit up what looked like a large footprint. Sure enough, there was a single footprint. We measured it to be fourteen inches long and, even more interesting, it was barefoot. Due to the location of the cemetery and the terrain that makes up the gravel road that leads to the cemetery, it is very unlikely that anyone would be walking barefoot in the area. Chris took a photo of the print and now has it on his Facebook page.

Ridge Cemetery at Williamsburg Hill

Before we left the cemetery that night, Stan left two sophisticated audio recorders, in the timber, near the graveyard. He uses heavy, military grade, ammunition boxes, that he had customized to his needs, in which he places his audio recorders. He then sets up these boxes in strategic locations and leaves them for several days to record the sounds of animals in the forest.

Stan left the recorders for several days and later retrieved them to review the audio. After he began the review, he heard something that, to this day, he cannot figure out. In order to open the lid on the boxes, it takes two hands, because you have to pull upward on the latch of the box with one hand and then open the box with the other. Stan's recorders are so sophisticated that they can record sounds thousands of feet away. What Stan recorded was the sound of the latch on one of the boxes being lifted up and then the lid being opened. Moments later, it was closed again. What is so odd about this is that there were no sounds of footsteps of either animals or humans and there were no sounds of any vehicles driving up in the parking lot. Only the sound of the lid on the box being opened could be heard. Stan, a very logical and analytical person, was completely baffled at how anything could get close enough to the box to open it without those sounds being recorded.

One of the most unusual things that I have noticed about Ridge Cemetery in the over seventy-five trips that I have made while investigating the Hill and filming *Strange Williamsburg Hill* is that not one time have I ever seen any animal in the cemetery. There are trees with acorns and nuts in them, but no animals. There are birds that fly around and occasionally will land in trees in the cemetery. There are also bats that will fly around at night, but never any type of mammal. I have seen deer tracks along the road leading up to the cemetery, but have never seen a deer or deer tracks near the cemetery. One lady told me that she has been coming out to Ridge Cemetery for over thirty years and she has never seen an animal in the cemetery. Every person that I interviewed for my documentary told me the same thing when I asked them about the animals; they all told me that they had never seen any animal in Ridge Cemetery. With the way animals seem to be able to sense supernatural and paranormal things, I find this quite interesting.

Investigation Summary

So what makes Williamsburg Hill so strange, and why has it been a source of so many legends over the years?

Is it simply the product of rumors and overactive imaginations? Do our minds play tricks on us, or do people actually come into contact with the dead either by some chance encounter or by laws of nature that we do not understand? If ghosts exist and hauntings occur, why are some places haunted while others seem normal? Why do some cemeteries feel peaceful, but others, such as Ridge Cemetery, give you the uneasy feeling of being watched or followed? Do portals and windows exist that allow travel between different dimensions or worlds, and if so, who is haunting whom?

With all of the things I have experienced that I cannot explain, all the incredible eyewitness accounts, and with all of the oddities that exist on that hill, I have to take note.

There is something strange taking place at Williamsburg Hill, something that has been taking place for years. I get the sense that there may be some type of cosmic consciousness at work there. Williamsburg Hill is like no other place that I have ever investigated. It is a special place, which allows you to step back in time. Something about the mystique of the area keeps me coming back for more.

The next time your daily journey takes you near Tower Hill, exit and turn off the main road. Head down Route 1100 East and check out Ridge Cemetery for yourself. You never know who or what you might encounter in this odd, but special place.

The Legacy Theater

CHAPTER THREE

The Legacy Theater

Springfield, Illinois

The rarest experience of mankind is the adventure into the world of the unknown, the world of the nonphysical: the supernatural. Those heavy footsteps heard in the hallway when you know that you are alone. Shadows that seem to dart past, but with the turn of the head they disappear. That icy chill on a warm summer's night, that causes the hair to stand up on the back of your neck; and the feeling of being watched or followed when no one is there.

What you are about to read is true and a matter of personal record. We cannot explain it nor can we prove or disprove it, but to the staff of the Springfield Theater Center and those who frequent it, there is no proof necessary, because they have experienced these unusual things firsthand.

Located in the heart of Illinois, the city of Springfield serves as the hub of the state's political activity. In 1839, Springfield became the state capital with the help of a young lawyer named Abraham Lincoln. Lincoln lived there until 1861, when he left to become the 16th President of the United States. From that moment on, the city's legacy was tied to this great American.

Many believe that some of the citizens from these bygone days still roam the streets and buildings of Springfield; citizens of a more spectral nature, who appear out of nowhere to haunt those whose paths they cross. The Legacy Theater and Springfield Theatre Center, located at 101 East Lawrence Street, is such a place.

Housed in an unassuming location, the quaint and friendly atmosphere of the theater gives you the feeling that you have taken a step back in time. Located not far from the State Capitol Complex, the Springfield Theatre Center, as it was formerly called, was the venue used to showcase live performances for the Springfield Theatre Guild.

On November 8th, 1951, the Theatre Center opened its doors with a performance of the Broadway show *Born Yesterday*. The Center received numerous congratulatory telegrams from many famous celebrities, such as Bob Hope, Bing Crosby, and Broderick Crawford.

One of the members of the Theatre Guild was an actor named Joe Neville. According to some, Joe was a little odd and at times arrogant. Some of the other cast and crew were not fond of him. Joe also had a mysterious side to him as well. It was rumored that he previously acted in England under a different name, but being a talented and dedicated actor, his past was overlooked by his fellow actors.

In 1955, Joe was given the lead role in the play, *Mr. Barry's Etchings*. Things were not as they seemed. After a dress rehearsal for the opening of the play, Joe returned home and committed suicide, by overdosing on pills. He never performed in the show. It was later determined that the reason Joe ended his life was due to an audit at his place of employment. Funds had been misappropriated and Joe was the likely suspect.

But as they say "the show must go on!"

After Joe's death, his role was assigned to another actor only one day before the show was set to open.

Even after his death, many of those who frequent the Legacy believe that Joe's spirit continues to linger inside the theater. Reports of paranormal activity began almost immediately following Joe's demise and continue to this day. Actors and stage crew have reported strange sounds; such as doors opening and closing on their own, lights turning off and on without reason, along with costumes and tools that disappeared only to be found later, folded or placed in an area which had been previously searched.

Many claim to have seen Joe's spirit wandering the Theatre Center, prior to the opening of a new show. Joe was known to use large amounts of *Noxzema* cream for a skin condition on his legs, so whenever Joe was around, the pungent odor would annoy many of the other actors and staff. After Joe's death in 1955, the use of Noxzema cream was banned from the theater. However, some still claim to smell the odor of Noxzema in the old dressing room.

Today, the Theatre Center has a new owner and a new name: The Legacy Theater. Owner Scott Richardson has started a new era as he greets the faces of fresh guests. The days of Joe Neville and the patrons of years gone by are simply memories of a faded past that are gone forever, or are they?

Strange encounters have led some to believe that spirits from the past continue to linger, refusing to move on to their final resting place. Preferring to remain or frequent the quaint and friendly atmosphere of The Legacy. Scott Richardson himself has experienced some of the strange phenomena that occur on a regular basis.

I first met Scott in early June 2011, just weeks after he purchased the old theater. I had been trying to get into the Theatre Center for months, but had been unable to find the right person who could authorize an investigation. Fortunately, my good friend Lynn Puls, a local hairstylist and owner of The Hair Shanty, was asked to help style wigs for a play taking place that July at The Legacy. Although Lynn had never met Scott before, he asked her to meet with him to discuss hairstyles for the play. During the meeting with Scott, Lynn mentioned that a friend of hers was interested in conducting a paranormal investigation of the theater. Lynn asked Scott if he knew who could be contacted to allow access to the building to do the investigation and Scott replied, "As of a few days ago, me!" As they say, the rest is history. I, along with another paranormal investigator named Jay Nandi, met with Scott during the first week of June and he gave us a ninety minute tour of the theater. During the tour, Scott explained some of the strange occurrences that he had encountered since he came into ownership of the building.

Scott explained to us how, at the time he was considering purchasing the theater, he had gone on several walkthroughs of the building. During each of the visits, he noticed that the building gave off a depressed or negative feeling. But, after he purchased it, and walked in the door for the first time, as the new owner, he noticed that it was giving off a completely different vibe. It had a feeling of relief, as if the building knew that it was about to get a face lift and soon would come alive again with performances and spectators.

The theater had sat idle for several years and needed a lot of work in order to get it in a condition where it could again host performances. There was so much work to do that Scott said he was not sure what needed to be done first. He decided that the first thing he would work on was the landscaping, to give the outside a fresh look.

Once that decision had been made, he did not know where to begin, so one afternoon he contacted a local landscaper and they arranged to meet the following day. Scott locked up the building before going home for the evening and took the only key with him. The next morning, when Scott unlocked and entered the theater, he found original building plans rolled up on his worktable. These were not just any plans; they were the original landscaping plans. Scott told me that he had been through all the paperwork that was left in the building, and the original plans for the exterior were not included with the paperwork. No one, but him, had access to the building. So where these plans came from, is still a mystery.

Strange things continued. One day, Scott was doing some repair work at the theater using a hot glue gun. He laid the glue gun down next to him and turned away for just a moment. When he turned back, the glue gun was gone and was nowhere to be found. Later, as Scott was getting ready to leave the building for the night, he noticed something behind the last row of seats in the building. Underneath the seat, and with its cord neatly wrapped around it, was the glue gun. Scott has no idea how it got there.

Another evening, Scott was alone and working in the building when he heard what sounded like someone taking a handful of nails and throwing them in the air on stage. He heard the pinging sound of nails landing on the stage, but when he checked it out, nothing was there.

Stage at Legacy Theater

On June 17th, 2011, Jay and I conducted an investigation at The Legacy Theater. It was an overcast night and the moon was in a waning, gibbous phase, which means it was ninety-seven percent full. We had a few showers and thunderstorms move through the area as we were conducting our investigation, which helped add to the unusual atmosphere.

We arrived at the Legacy around 7:00 p.m. and the owner arrived a short time later. Scott decided to stay for a while and do some spackling work in order to prepare the walls for painting. As Scott went about his business, Jay and I did our walkthrough. We recorded baseline readings for temperature and electromagnetic field (EMF) readings. The difficult thing about that particular investigation, as far as EMF readings go, was that there are a lot of electrical wires running underneath and around the stage area, so any fluctuations in EMF in this area were attributed to all of those utility wires. The temperature was, for the most part, about seventy-five degrees upstairs and slightly cooler in the basement.

Scott left around 9:00 p.m., and at that time Jay and I set up our video cameras and strategically placed our digital audio recorders throughout the building. For the most part, things were very quiet and we did not have any personal experiences. At one point, however, around 11:00 p.m., I was standing in the doorway of one of the old dressing rooms that Joe Neville would have used. Jay was in the dressing room and was conducting an EVP session. All of a sudden, I felt someone or something tickling my back. My first thought was that a spirit was touching me, but as I turned around, I discovered that a bat was fluttering on my back!

As the night progressed, we were somewhat disappointed that we were not experiencing any activity. After all, a place like the Legacy has had many reported paranormal experiences over the years by credible people who did not want or expect to interact with the dead.

Around 4:00 a.m., we completed our investigation and began to break down our gear. With hours and hours of recorded audio and video to go through, we were in hopes that our time and effort would be fruitful. As it turned out, it was. Almost immediately upon reviewing our evidence, Jay and I found audio that proved we were not alone during our investigation, and that we had more company than just a pesky bat!

The Legacy Theater produced some of the best EVP evidence that I have recorded in the eight years that I have been doing paranormal research. It is important to keep in mind that Jay and I were the only two in the building. I rarely investigate with more than three or four total investigators, in order to keep the environment controlled and to make reviewing audio less complicated. It is much easier to determine what is said, and by whom, when there are only one or two people present during a recording session.

One of the strangest EVPs was one that seemed to be of someone speaking with a foreign accent. In this particular clip, Jay and I are having a conversation about the investigation. Although Jay was born and raised in central Illinois, and therefore does not have an accent, his heritage is from India. Right after Jay spoke; you can hear a loud and clear voice with an Indian accent say what phonetically sounds like *Creloza*. We do not know what it means or if it is possibly a person's name. I have talked to people of Indian decent and they have never heard the word before. We have also tried using online language converters without success. This may be one of those clips that will remain a mystery.

The next EVP also seemed to involve Jay. In that clip, Jay was talking and can be heard saying, "At the same time..." Then, immediately, a voice that we did not physically hear, but that was recorded using a digital voice recorder says, "fruit cake," as if to make fun of what Jay was saying.

During the investigation, Jay and I were down in what was the old concession area near a storage room. I had placed my digital recorder near the door on a metal folding chair. A sump pump started running, so I decided to close the door to muffle the noise. I noticed that an extension cord was draped over the top of the door, so in order to close it; I first had to move the extension cord. I was carrying a video camera with me, so I laid it on the chair next to the audio recorder. When I reviewed the audio, I heard the sound of the cord moving, but then, I clearly heard a male voice whisper, "You left shit there," as if to remind me that I had laid my equipment on the chair. Or, maybe, I was invading the space of the spirit by laying my equipment on the chair. This was one of the clearest EVPs that I have ever recorded. The spirit seemed to be interacting with me, which would make this an intelligent and not a residual haunting.

In case you are not familiar with the different types of hauntings, a *residual haunting* is like a recording or an imprint in time; somehow the energy of the spirit was imprinted on the environment in which they were living. An *intelligent haunting* is a haunting in which the spirit or entity can interact with the living in the physical world and seems to have purpose to it.

We recorded a fourth EVP that also seems to be intelligent. In this EVP, you can hear me saying to Jay how I was a little disappointed that things seemed pretty calm and that we had not had any personal experiences. At the end of the clip, Jay jokingly said "Well, at least we experienced a bat flying around." In between what Jay and I are saying, you can clearly hear a male voice say, "That's your own conclusion." It seemed that a spirit had a difference of opinion!

Probably the most amazing EVPs of the night were recorded in the area around the old dressing room. On two separate occasions, we recorded singing about forty-five minutes apart. This was not just ordinary singing, but what sounded like a man and a woman rehearsing for a play or performance. It is simply amazing. All in all, we recorded six EVPs that I classify as *Class A*, and four other good quality EVPs that needed to have the volume enhanced to hear them better.

Due to all of the audio evidence that we obtained, I plan on conducting a second investigation with my Urban Paranormal team. Hopefully, whatever or whoever haunts the theater will be even more willing to interact with us our second time around.

Investigation Summary

The investigation at the Legacy Theatre was not an investigation where we encountered anything personally, but there have been many people over the years that have had personal encounters and experiences there that they cannot explain. Jay and I had an encounter the night of June 17th, 2011, but it was one that we did not realize until we reviewed our evidence. It became apparent that we were not alone in the building and that we encountered the world of the unknown. Fortunately, modern technology allows us to experience this world, which cannot always be seen with the naked eye or heard with our ears. It is an intangible world, but a world that exists nonetheless, the world of the supernatural.

Based on the clear audio evidence that we recorded, coupled with what appears to have been some kind of intelligent interaction with both of us. I can conclude that there is activity-taking place at the Legacy Theater. To me, this lends credibility to the stories that have been told over the years.

Who was interacting with us? Was it Joe Neville, or was it one of the thespians or patrons from days gone by who still roam those halls? Hopefully, with the use of our equipment, and with continued advancements in ghost hunting technology, future investigations will provide even more evidence and clues to who the theater is being haunted, and why.

I have investigated many places over the years, and the more that I investigate and delve into the supernatural, the more I get a feeling about certain places. It is hard to explain, but in some locations this extra sense kicks in and you pick up a vibe or a feeling that seems to overwhelm you. When this takes place, you have a feeling that something is not right, that you are being watched or followed. The whole atmosphere seems to change. Many times, a feeling that you are not welcome follows this experience. For whatever reason, the air seems heavier and gives off an entirely negative feeling. The Villisca House and Williamsburg Hill are two places where I have had this feeling, but I have had it at other places as well. Then there are places like the Rockcliffe Mansion, where this extra sense kicked in. In that particular place, that heavy feeling remains in the air and you become aware of being watched, or at times followed, but you are not alarmed, because that feeling tells you that whatever is there simply decided to stick around after death and continue to reside or frequent the place that they lived, worked, or loved in the past.

The Legacy Theater gives you that type of a feeling. Whatever is there makes you feel comfortable and welcome. It is a very positive feeling. I agree with Scott Richardson, the owner, who had that feeling, the first time he walked into the theater, after he bought it. It felt as though the building gave a sigh of relief that someone with such a great passion for the arts, as well as someone whose intension was to preserve and protect that wonderful theatrical landmark, had purchased it.

The next time you pass by The Legacy Theater at 101 East Lawrence Ave in Springfield, stop by and take in one of the performances. Feel free to strike up a conversation with one of the friendly patrons, but if you start to smell the pungent odor of Noxzema, or you feel a cold chill and the hair begins to stand up on the back of your neck, you may want to take a second glance at the person sitting next to you... just in case!

Rockcliffe Mansion

CHAPTER FOUR

Rockcliffe Mansion

Hannibal, Missouri

A few years ago, I was fortunate to have the opportunity to do an investigation at Rockcliffe Mansion, with my good friend, Dr. Gary Hawkins, and a gentleman named Paul Robinson. At the time, Paul worked for the local FOX TV station in Columbia, Missouri. I had previously worked with both of them on a paranormal investigation at the former location for the LaBinnah Bistro, which is also in Hannibal, Missouri.

Paul was in the process of putting a documentary together called *Haunted Hannibal* and had been utilizing Dr. Hawkins to assist him in the venture. Gary knew of my interest in learning more about conducting paranormal investigations, so he was kind enough to invite me to assist in the LaBinnah Bistro and Rockcliffe Mansion investigations. I gladly accepted the opportunity to work with someone as respected in the field of paranormal investigation as Gary.

Not only is Gary an excellent paranormal investigator with psychic abilities, but he has a unique ability that no one else possess, at least that I am aware of. Gary claims to have the ability to catch ghosts. Yes, I said *catch*. I know this is hard to believe, and when Gary first told me about this unique ability of his, I was, to say the least, more than a little bit skeptical. But after seeing Gary in action at the LaBinnah Bistro, and having the opportunity to actually feel the energy that he had captured simply holding out his right hand, I quickly became a believer. His hands act like a magnet, latching onto energy, be it ghostly or otherwise. When Gary catches a ghost, or energy, he always allows others who are with him to feel the energy that he has captured. One thing that I have learned in the years that I have been investigating the paranormal and supernatural is that many things happen that we cannot explain. Gary's ability to *catch* energy is one of those things for which I have no explanation.

The Rockcliffe Mansion was built in 1900 by wealthy lumber baron John Cruikshank. Cruikshank wanted a house that would showcase the finest woods and furnishings that money could buy. The mansion was built using the finest Walnut, Oak, and Mahogany available at the time. The exterior of the building is of double brick construction. Inside, you will find some of the original furnishings used by the Cruikshank family.

The Cruikshank's consisted of John, his wife, and four daughters. They lived in the mansion along with their butler until the death of John Cruikshank in 1924. After his death, the mansion was left vacant for over forty years. It was weeks away from being torn down, when a local group got together at the last minute and purchased the old mansion, restoring it back to its original form. Today, tours are conducted at Rockcliffe Mansion and it is also a seasonal bed and breakfast. The mansion is enormous, spanning 13,300 square feet, including nine bedrooms and seven bathrooms.

Rockcliffe Mansion is believed to be haunted by the spirits of John Cruikshank, his wife, four children, and their butler. The owner at the time of our investigation claimed that he once saw a five feet, four-inch impression of a figure in the bed sheets on one of the beds. Many guests claim to smell cigar smoke in one of the rooms in which Mark Twain once stayed. Twain gave his last speech in Hannibal in 1902 and stood on the third step of the main staircase as he addressed some three hundred invited guests. There are reports of camera crews that have come there to film and their batteries will go dead or their cameras will malfunction.

On the day of the investigation, Gary and I arrived at the mansion at 2:00 p.m. Paul arrived a short time later. The plan was for Paul to start conducting interviews and videotaping eyewitness accounts of people who had experienced paranormal activity at the mansion. Paul had also advertised for several people to attend the investigation who were interested in ghosts and hauntings, but had never been on a ghost hunt. The reason for this was so that he could film their candid reactions to any activity that may have occurred during our investigation. Over forty people applied for the opportunity to attend and Paul selected three of them. They were due to arrive between 5:30 p.m. and 6:00 p.m.

While Paul was busy setting up for the interviews, Gary gave me a quick tour of the mansion. As we were walking around, a young lady who works there stopped us.

She wanted to let us know that she was leaving, but would be back a little before 7:00 p.m. to finish setting up decorations in the basement. From 8:00 p.m. until 10:00 p.m. She was hosting a small bachelorette party for a close friend of hers, and she stressed that they would keep the noise down, because she knew we were conducting an investigation. Gary let her know that this would not be a problem, as we would investigate the basement after the party was over.

The young lady told us about an experience she had only a few weeks before. She was working in the mansion one evening and was sitting at a desk near the main entrance. "It was raining outside, of course," she laughed. The doors were locked and she was alone in the mansion. Suddenly, she heard footsteps directly above her.

You didn't hear that, she thought. Trying to convince herself, that she only imagined hearing the noise.

Moments later, she heard the footsteps again, and a few minutes after that she heard the footsteps a third time. The third time, she stood up and walked to the foot of the staircase and yelled upstairs, "If you are trying to scare me you are not doing a very good job!" After confronting the noise, it stopped.

Around 5:00 p.m., Gary decided to take a walk around the grounds of the mansion, so I left him and accompanied Paul, who was getting footage of the mansion in the daylight. We were coming down the main staircase, from the second floor, when Paul saw a shadow to his left, in an area on the first floor, that lead into the dining room. Due to the angle of the staircase, which went from the second to the third floor, there was only limited view of the dining area from where we stood. Paul described the shadow as completely blocking out the entire portion of wallpaper, just above where the wood wainscoting and the wallpaper met, near the entrance to the dining area. He and I were starting to get a little excited, both feeling that if things were starting to happen this early in the evening, then it might be a very eventful night.

Around 6:00 p.m., a couple from St. Louis named Kimberly and Tony arrived, and Chuck, a gentleman from Hannibal, followed them shortly thereafter. All three had responded to Paul's advertisement. Upon their arrival, we took them into the main dining room where Gary and Paul were going to give our new friends a brief rundown on what to expect during the evening, including how the investigation would be conducted. Gary usually gives what he calls his *Ghost 101* speech to those who are not familiar with paranormal investigations, which includes a rundown on what ghosts are and the forms in which they appear. He also reassures his audience that ghost cannot harm them and that there is nothing to fear from ghosts.

With the exception of Gary, who stood at the head of the table with his back to the main living area, the rest of the group was seated at the table. On one side of the table, with our backs to the entranceway, were Kimberly and myself. Kimberly's boyfriend, Tony, sat at the far end of the table, seated next to Paul. On the other side of the table, seated across from me, was Chuck. Gary gave his Ghost 101 speech to the three new ghost hunters, as well as discussing our plans for dinner. Normally, after the mansion closes for the day, the owner locks all of the doors and sets the alarm system so no one can enter without being detected. In order for us to have access afterhours, the owner left only one door unlocked for us to go in and out. We were told that if we all left we would have to lock up and would not be able to re-enter. So, in order for us to go to dinner, our plan was to wait for the employee, who would be returning at 7:00 p.m., to finish setting up for the bachelorette party. We could let her in, then she could let us out and lock the door, so we could all leave to go eat. When we returned, she could then let us back in.

Interior shot Rockcliffe Mansion

About 6:30 p.m., Gary was finishing up his talk when I noticed that Tony had a puzzled look on his face. His expression was almost like a frown. Finally, Tony said, "You guys may think that I'm crazy, but I keep seeing shadows through the doorway moving across the living room." Paul then explained to the group how he had seen a shadow earlier as he was coming down the stairs about an hour or so earlier. A few moments later, I noticed a similar expression of puzzlement on Chuck's face just before he exclaimed, "Tony, you're not crazy, I keep seeing a shadow moving across the wall in the living room myself!"

With people starting to see activity this early in the day, we figured that we were in for an exciting evening.

At about ten minutes before 7:00 p.m., the young lady who was in charge of the bachelorette party still had not returned. I suggested to the group that, if they wanted to leave to go eat, I would stay behind and watch the place. They could bring back something for me to eat when they returned. Gary suggested that we give her a few more minutes and, if she did not return by 7:00 p.m., the group would do as I suggested.

At almost 7:00 p.m., Kimberly, Tony, and I all turned, looking toward the side door to which the owner had allowed us access. We had locked it from the inside, but we heard what sounded like someone unlocking the door, shutting it hard, and then dragging something across the wooden floor. As soon as I heard the noise, I said to Gary, "The girl is here, let's go eat," and he eagerly agreed.

We all got up from the table and walked into the main downstairs area, then headed toward the door leading to the basement. Gary yelled downstairs to the basement where the party was to be held, since he wanted to let the lady know that we were heading to dinner and that she would need to let us back in when we returned. No one answered. He yelled a second time, but again no one answered. Gary and I looked at each other and then proceeded down the steps. We looked around the basement, but there was no one around. We then searched the entire mansion and discovered that no one was there, other than the six of us. We checked the door and it was still locked from the inside. Kimberly, Tony, and I all reassured each other as to what we had just heard, but we had no explanation as to what caused the strange sounds.

That was not the end of the unexplained occurrences that night. Since the young lady had not shown up yet and it was approaching 7:00 p.m., we decided that I would stay to keep an eye on the place while the group headed out for dinner. We were all gathered beneath the chandelier in the main foyer, discussing dinner plans, when several of us heard a metallic tapping sound. It sounded as though someone was tapping with their fingernails on the chandler that was directly above our heads.

When the group left, I locked the door behind them and grabbed my digital recorder and camera to explore the mansion, in hopes of capturing some solid evidence of a haunting. All was quiet as I walked through each and every room on the main floor. I did hear another tapping sound, kind of a *tap-tap-tap* that would stop and start again. I could not determine where the sound was coming from, other than it was not coming from the main floor. I went up the main staircase to the second floor and walked through each room, determining that the sounds were not coming from the second floor either. I continued upstairs, to the third floor. Again I searched all of the rooms trying to identify the source of the sound, but again I found nothing.

The only room on the third floor that gave me an odd feeling was the ballroom, which had also been used as a playroom by the Cruikshank children. The room's walls were painted sky blue and the woodwork trimmed in white. The restored wooden floor had a dull look to it as though it was still being used. With every step I took, the floor creaked, just like in the old movie *Bowery Boys,* when they were in a haunted house. The room felt as though it was occupied.

Since the tapping noise was not coming from the third floor, the only other possible place it could be coming from was, of course, the basement.

This time, I used the staircase on the west side of the mansion, and as I neared the first floor I could hear the tapping sound again. The question I now had to face was whether I wanted to go down in the basement alone, in a reputedly haunted mansion, to check out a strange tapping sound. My investigative nature overtook my lack of better judgment and I decided to continue down to the basement. The basement had several rooms, along with a stage area, which was used to host cabaret type shows and other functions. I heard the tapping sound coming from what I later determined to be some type of kitchen or food preparation area. As I walked toward the noise, I realized that a pipe used in the heating of the house most likely caused the tapping. It was disappointing, to say the least.

I went back upstairs and camped out in the dining area. I sat down where Tony had been seated earlier in the evening in hopes to catch a glimpse of the shadows that Chuck and he had seen. I was alone in the house for over an hour when one of the ladies finally showed up to finish the decorations for the party.

During the hour and a half the group was out to dinner, no activity took place. I was getting the feeling that whatever haunted this place had been feeding off the energy from the group. Considering that there were six people around and lots of camera equipment before the group went to dinner, there had been plenty of energy for the spirits to draw from. So, with just one person in the mansion, there may not have been enough natural energy for ghostly things to occur.

About 8:30 p.m., the group returned and it was time to get started with the investigation. The plan for the night was for us to set up one of Gary's infrared surveillance cameras in a downstairs room that had been a favorite of Mrs. Cruikshank when she was alive. The story was that Mrs. Cruikshank loved to play the grand piano there. A second surveillance camera would be set up on the third floor in what used to be the ballroom where the Cruikshank's hosted

CHASING SHADOWS

parties and get-togethers. That room was also used as a play area for the children.

Modern day guests to the mansion have claimed to hear the sounds of small footsteps walking and running throughout the mansion. Paul Robinson told me that during a previous investigation, he was standing on the second floor just below where the ballroom is located and heard the sound of small feet running across the floor. He said that they tried to recreate the sound by having one of their group members run across the floor, while he stood in the same spot, on the second floor, where he had heard the sound. Paul said that only when the person ran on his tiptoes did it sound like the noise that he had heard. He believed that it was the children who had made the footsteps.

Once the surveillance cameras were set up, we were ready to start the investigation. Gary provided a laser pointed digital thermometer to Tony, and his EMF detector to Kimberly, to use with the investigation. The idea of the thermometer is to check for temperature fluctuations. When ghosts are present they need energy, and when they draw energy from natural sources, such as people or electronic equipment like cameras and flashlights or from natural magnetic fields, they can cause cold spots. Investigators often use EMF detectors to determine whether or not there are any electromagnetic disturbances in the area.

It is believed that EMF detectors pick up on spirit activity. The only pitfall of an EMF detector is that it also picks up on the electromagnetic fields created by natural sources. Thus, an investigator is required to do an initial walkthrough of any location that they investigate, in order to make note of any naturally caused electromagnetic field sources.

Several times during the afternoon, when Gary and I were in the room in which Mrs. Cruikshank use to enjoy playing her piano, Gary had a feeling that there was a female presence in the room. So when we were ready to begin our investigation of the first floor, and in particular this room, he suggested that since there are two doorways leading into the room, that he would go in one doorway and the rest of the group, would use the other. He felt that by entering in this manner, if there was a ghost in the room it would have to pass by, or through, one of the group in order to leave.

Gary entered through the doorway on the right, while Paul, who was filming, and the other three members of the group entered through the doorway on the left. I followed Paul with my camera.

No sooner had Paul passed through the doorway when I heard Gary yell, "I've got one!" I decided to turn back and film Gary to see if I could catch anything on video. I was still filming as I was walking toward the doorway. When I reviewed my video footage the next day, a child's voice saying "Momma" can be heard at the point where I am walking toward the doorway, as though it was wondering where its mother was.

When I got to the doorway where I could see Gary, he had his right hand extended, palm up, and he told us that he had immobilized a ghost. Unless you have ever experienced Gary catching a ghost, or whatever it is that he latches on to, and have actually felt the energy that he captures simply by holding out his right hand, I would not blame you if you did not believe it could be done. What I do know is that he catches *something*, whether it is his own energy or that of a ghost.

I filmed Gary as he gave each of the three first-time ghost hunters a chance to feel the ghost he had captured. When Gary gave his speech earlier in the evening, he explained his unique talent of ghost catching. I could tell by their expressions at the time that they were very skeptical of Gary's claim, but what I caught on film was their skepticism changing to belief. Gary could do what he claimed.

As Gary gave each of them a chance to experience firsthand what he had captured by placing their hands in a similar position as his, I could tell by their verbal and physical expressions that they were amazed. Gary also gave Paul and I the opportunity to feel the ghost. What I felt was a tingling resistance that caused my hand to feel as though it was going to sleep. Paul said it was, "Like you could reach out and grab a hand full of something." Whatever Gary caught, whether it was a ghost or some type of natural energy field, he was holding something like a magnet to metal.

What happened next was even more bizarre. While Paul was filming Gary catching the ghost, he said, "Gary, I caught a nice orb near you." Once Gary released the paranormal energy that he had captured and the excitement had subsided, we decided to take a short break. During the break, Paul asked me if I would mind filming the reactions of the three new ghost hunters, while he replayed the footage of the orb that he had captured near Gary. I agreed to do so.

Paul showed the orb that he had captured on film to Gary and the rest of the group. What could be seen was a small ball of light floating from about five or six feet in front of Gary to almost his knees before reversing direction.

The group was pretty excited over this when Gary pointed out something else during the footage. Paul rewound it and played it again. The reactions of several members of the group were priceless. When Chuck saw the footage he was so in disbelief at what he had just seen that he walked out of the room shaking his head.

First, the orb floated toward Gary and then reversed direction. A few seconds later, with Gary standing completely still, a shadow, or more like a silhouette of a man, came from behind him, walked the opposite way that he was facing, and then out of the room.

I was shocked when I saw this. There was no source of light to cause the shadow. The only light was coming from outside. If that had been the light source, the shadow should have been in front of Gary and not behind him. Everyone, with the exception of Gary and I, were standing next to Paul. I was standing just outside the doorway and to Gary's left. I was standing completely still. So everyone was accounted for, all sources of light were accounted for, and there was no logical explanation for the silhouette. It was as though Gary's own shadow walked away from him without Gary moving.

The three amateur ghost hunters, two who seemed skeptical of ghosts and hauntings before our investigation began, now seemed at least puzzled. They were searching for logical explanations for the shadow that they had just seen in Paul's video footage, but try as they may, they could not come up with any. They knew that everyone was accounted for and that no one else was in the mansion at the time the shadow was recorded on video.

They also could not offer an explanation as to what they felt and experienced when Gary gave them the opportunity to feel the ghost that he had immobilized. They knew they had felt some type of energy field, be it of a ghostly nature or some other unexplained phenomena. Chuck, one of the amateur ghost hunters, was a little shook up by what he had experienced. Not that he was afraid, but it was the fact that he had experienced things that are not normally experienced and that he could not logically explain. That is what investigating and experiencing the paranormal is all about.

After the excitement pertaining to the shadow died down, we proceeded to the second floor, but did not encounter any activity there. We then proceeded to third floor ballroom to continue our investigation and it proved to be just as interesting as the room on the first floor.

The ballroom was primarily an empty room with a few items still stored in it. In one portion of the room there were several mannequins dressed in period clothing and a small doll was also propped up near one of the mannequins, which gave the room the feeling that children still occupied it. As I mentioned earlier, the restored wooden floor had a dull look to it as though it was still being used as a playroom. We wanted to see if we could capture any evidence of the children playing.

Paul and I had our handheld video cameras running and we had already set up one of Gary's infrared surveillance cameras, which had been filming while we were doing our investigation downstairs. I had also placed my digital voice recorder on a small end table in the room earlier in the evening in hopes of capturing any possible EVPs. The group spread out around the room and we all sat down on different sides and were leaning against the walls in hopes of either hearing or seeing some ghostly activity.

After we had been in the room for some fifteen minutes, I thought I heard the sound of a child murmuring. I was not sure, so I was not going to say anything, but then Paul asked, "Did you hear that?" and I replied, "I did." "It sounded like a murmuring," Paul continued. I then said, "It sounded like a child." Paul agreed. On video footage, the murmuring can be heard just before Paul says "did you hear that?" The murmuring is actually present three separate times on my handheld video camera. All three times that the murmuring was heard, it was acknowledged as being heard by a member of the group.

While the group was still sitting around the third floor ballroom, Gary decided to walk around the room to see if he could either feel any cold spots or get a sense of any activity. He passed roughly eight feet to my right. He said something brushed up against him and that the air around him felt very cold. At the same time, neither Paul nor I could get our video cameras to focus on Gary. It was like something was between us and Gary. Our cameras were trying to focus on something that was not there.

Several minutes passed and Gary said the cold feeling had gone away. Almost on cue, our cameras began to focus again. Gary continued to walk around the room, when I thought I saw a shadow to my left and all the way across to the other side. I asked Gary if he would check it out for me, and he walked to the area where I thought I had seen the shadow. When he walked to the spot, Gary said, "You saw something, I can feel it, and it's extremely cold here."

As Gary was standing there, he placed his hands in his pockets. Just a few seconds after he did this, I saw the shadow of an arm and hand move up and down. It was just to the right of Gary and below his hip. I immediately asked, "Gary, did you move your hand?" At the same time, Paul said, "Did you hear that?" When I replayed my video footage, what I captured was the shadow of an arm and hand moving up and down on the wall behind Gary. As Gary moved slightly forward, the wall behind him can be seen and it is somewhat lit up by the beam from my camera. All of a sudden, the lighted spot disappeared and a shadow filled in behind Gary. The strangest thing about this was that Gary was standing perfectly still when the shadow blocked out the light that was behind him. Upon further review of my video footage, I noticed that the murmur can be heard again just a second before Paul asked, "did you hear that?"

Many times, EVPs are captured on tape or video, but are not heard by the naked ear. This time, we had personally heard the sounds as well as captured them both on video and the digital voice recorder. When I reviewed my digital voice recorder, I found that, in addition to the murmurs, I had captured several sounds when no one was in the room. These sounded like someone was holding a mop by its handle and letting go so that the handle fell and hit the wooden floor. I recorded this sound twice. No one was in the room when these were recorded. Were the noises that were captured on audio and on video made by the ghost of the Cruikshank children? Unfortunately, I have no explanation for the strange sounds.

The next stop on our investigation of the Rockcliffe Mansion was the basement. The basement had several rooms and an area that was used to host cabaret type shows and was formerly the location of the LaBinnah Bistro Restaurant, which is a seasonal restaurant and was not open at the time of our investigation. Would the basement prove to be as active as the first and third floors?

The basement was not as exciting as I had hoped, although Paul did capture some very unique orb footage. In one corner of the basement, Paul recorded footage of what I call a *Star Wars* type light show of orbs. The orbs were all rapidly moving about and seemed to be directed toward

Paul. There were at least twelve to fifteen orbs zooming about in Paul's direction. Again, there was no explanation for this activity.

While in the basement, Chuck and I were standing near the stairs leading from the basement to the first floor when we heard heavy footsteps and then the sound of someone running across the floor above us. At the time we heard the footsteps, the entire group was in the basement. Chuck and I went up to the first floor to take a look around and, of course, no one was there.

At about 2:00 a.m., we decided to call it a night and pack up the gear. We said our goodbyes to Kimberly, Tony, and Chuck. When they had left, Paul, Gary, and I finished packing up. I decided to use the bathroom before making the two hour trip back to my home in Taylorville.

When I went into the first floor bathroom just off of the kitchen area, it was dark and I could not find the light switch. I returned to the kitchen were my equipment was stored and retrieved a flashlight. I returned to the bathroom and when I pressed the button to turn my flashlight on, the bulb made a popping sound, followed by a flash of light. Something caused the bulb in my flashlight to blow, and it cracked the glass in the light! The flashlight was a few months old, so the bulb should have been in good shape. Even at that, I have had batteries go dead in flashlights, but never in my life have

I had a bulb blow like that. Was this paranormal or just coincidence? With all the things that happened this night, it made me wonder.

Investigation Summary

While I was investigating the Rockcliffe Mansion, all of the classic TV movies about ghosts and haunted houses kept flashing through my mind. If I was making a movie about a haunted house, this would be my choice for the perfect location. This place had rickety wooden floors that creaked with every step, sounds of unexplained footsteps, mysterious tapping on the chandelier, and even shadows that moved when no one was present. Add to that the unexplained voice of a child saying "momma", coupled with the murmuring sounds that echoed through the old ballroom on the third floor, and the only thing that was missing was a secret passageway that opened up by pulling out a book from the bookcase. This place is a ghost hunter's dream.

I have never been under the illusion that I have any psychic abilities, but all throughout the twelve and one half hours that I spent in Rockcliffe Mansion, I had the feeling that the Cruikshank family was still living there.

Several of the rooms that I went into alone felt as though someone was with me, watching my every move. It was not a creepy feeling, but on occasion I felt as though I was intruding on someone's privacy, as though I had interrupted what they had been doing. I had this feeling in both the first floor room just to the right of the main entrance where Mrs. Cruikshank's piano sat, and also in the third floor ballroom that the children had used as a playroom.

Is Rockcliffe Mansion haunted, you ask? Well, let us examine the evidence. What follows is just a sample of the many eyewitness accounts.

The first belongs to the young lady who works at the mansion and had an experience just a few weeks prior to our investigation. If you recall, the young lady described the experience of hearing heavy footsteps coming from the floor above her. She heard the footsteps on three occasions one evening while she was working late. When speaking with her, she seemed to be a very credible person.

The second bit of eyewitness evidence would be the moving shadows that Paul, Tony, and Chuck had seen.

The third piece of eyewitness evidence was the event that took place while our investigative team was sitting at the dining room table. Three people, including myself, distinctly heard the sound of a door being unlocked, opened, and then what sounded like a heavy object being dragged across the floor. We had thought it was the young lady returning to finish setting up for the bachelorette party. However, upon investigating, we found the door was still locked and that there was no one in the mansion, but our team.

The fourth piece of eyewitness evidence occurred while we were in the basement. Chuck and I heard the sound of someone running across the first floor directly above us. When we went up to check, no one was there.

The fifth piece of evidence took place on the third floor in the ballroom area. Several times during the evening, Paul and I heard the sound of murmuring in the tone of a child's voice. On one occasion, Paul asked if whoever made the sound would make another sound. Within a few seconds we heard the murmuring again as though it had been listening and was trying to communicate with us.

The sixth piece of evidence occurred several times during the evening when both Paul and Gary felt unexplained cold spots. On one occasion, Paul's arm became extremely cold while he was filming. At the same time, he could not get his camera to focus. My camera, which was focused on Paul, also became blurry and unable to focus.

Paul had over seventy minutes of battery power left on his camcorder when this occurred, but suddenly, without any logical explanation, the camera showed that the battery was low. Paul's digital camera batteries were also drained, so two independent sources of energy were drained simultaneously.

We were able to document much of the evidence that I just listed above on video and digital audio recorders, as well as some evidence that was not listed above. I was able to capture the sounds of the murmuring on both my digital recorder and on my camcorder. I also captured the sound of a child saying "Momma" while I was in the downstairs area of the mansion. This was recorded by my camcorder.

Both on the third floor and in the ballroom, my camcorder captured the shadow of a hand and arm as well as a shadow that moved behind Gary. This was also where I recorded the murmuring sounds. Paul captured several instances of floating orbs and one absolutely unbelievable video shot of a group of orbs flying around like something out of *Star Trek*.

The most compelling piece of evidence captured on film during the investigation was of a shadow that simply walked out from behind Gary and out of the room. It was unbelievable. Everyone in the mansion was accounted for at the time. Paul was on one side of the room videotaping Gary, while Kimberly, Tony, and Chuck were standing behind Paul. I was just outside the doorway on the same side of the room as Gary, filming him from a different angle. There was no source of light for the shadow other than an outside light that was behind both where Gary was standing and where the shadow appeared. If that light was bright enough to cause a shadow, which it was not, it would have cast Gary's shadow in front of him and not behind him. Plus, the shadow moved in the opposite direction of Gary's movements, as minimal as they were.

One last thing that I personally experienced was the bulb of my flashlight literally popping and making a bright flash when I turned it on in a dark bathroom. Not that this is proof of any paranormal activity, but how many times in your life have you switched on a flashlight and had this happen? Your answer, I will bet, is never.

The question now is whether I consider the Rockcliffe Mansion to be haunted. My answer is a definite, "Yes!" With all the credible stories, eyewitness accounts of previous experiences, and what I personally experienced with my own five senses, I absolutely believe that Rockcliffe Mansion is haunted.

LaBinnah Bistro

CHAPTER FIVE

LaBinnah Bistro

Hannibal, Missouri

My first investigation with Gary Hawkins was at the former location of the LaBinnah Bistro Restaurant at the corner of Center and North Fifth Street in historic Hannibal, Missouri. This was a very intriguing place to conduct an investigation due to the history of the building itself. The

Bistro is connected to Hannibal's famous unsolved Stillwell Murder case.

As the story goes, the former mayor of Hannibal, W.A. Munger, owned the building. Amos J. Stillwell, Munger, and the town's doctor, Dr. Joseph C. Hearne, were playing euchre at the mayor's house on the evening of December 30th, 1888. Rumor has it that the doctor and Stillwell's wife, Fanny, were having an affair. After the card game that evening, Stillwell and his wife returned home and went to bed. Sometime during the night, Fanny awoke to see a dark figure standing over the bed with an axe. The dark figure called out, "is that you, Fanny?" before swinging the axe and severing the head of Amos Stillwell.

Mrs. Stillwell was later seen in her nightgown, running around the neighborhood, knocking on doors seeking help. When the police arrived, Mrs. Stillwell and several of the neighbors had cleaned up the bloody scene. The next year, Fanny and the doctor were married. At that time, the Stillwell murder was still unsolved by authorities, but the gossip around town was that the doctor and Fanny were in cahoots. It was believed that they had murdered Amos Stillwell so that they could be together.

Most townsfolk were convinced that both Fanny and the Doctor were guilty of the murder. They would harass them with taunts, such as "Axe me no more questions!"

Eventually, the doctor went to court, but was acquitted by a jury consisting of several of his friends. In later years it was reported that during a family argument Dr. Hearne threatened Fanny with, you guessed it, an axe.

In more recent years, the Munger home became the location for the LaBinnah Bistro Restaurant and Social Club. The Bistro is located downstairs, while Hazel & Kirk's Coffee Bar is located upstairs.

At the time of our investigation, the Bistro's owner had relocated the restaurant to the Rockcliffe Mansion in Hannibal, which he also owned, and was in the process of selling the building where the Bistro was formerly located. He was aware that Gary Hawkins and Paul Robinson were interested in doing an investigation and shooting video footage, for a documentary that Paul was in the process of making called *Haunted Hannibal.*

Since the current owner was not sure whether the building's new owner would be as receptive to allowing a paranormal investigation, he quickly contacted Gary before the sale was complete.

The stories about the Bistro went back a number of years. An apartment above the second story of the restaurant had been, until recently, rented to a lady from New Orleans. One evening, she witnessed a translucent figure moving hurriedly across the dining area. She watched it turn into a glowing, blue dot before it disappeared. This frightened her so much that she moved out. Weeks later, a guest took a picture of the front of the building and in the picture, looking out the front door glass, was a figure similar to what the former tenant had seen.

Needless to say, when Gary asked me if I would like to go along on the investigation I eagerly accepted the invitation. It would be a chance to check out a place with such a rich history and lore. The investigation was scheduled for a Saturday night. I met up with Gary and Paul at a prearranged location and we drove to the site. This was my first time meeting Paul, who was originally from England and he had moved to the United States seven years prior. He has a wonderful British accent and was a real pleasure to work with. He now lives in Kansas City, Missouri and is involved with producing a sci-fi miniseries.

Meeting us at the location was local psychic Kae Blecha, her boyfriend, and Rick Rose, who owned the building at the time. Rick gave us free reign of the premises for the night. Kae is a local tour guide for Haunted Tours in Hannibal and also is a palm reader. She gave us a rundown on the activities that had been witnessed by the former tenant, including the sighting of the translucent figure. She also showed us a copy of the picture of the figure that had been peering out of the front door of the Bistro. The figure appeared to be almost childlike and was either very short or was possibly bent down looking out the door. A face could be made out, but no other real distinguishing features were apparent.

The home itself was a quaint; two story red brick structure with a limestone foundation and small front porch with limestone pillars. It has seen better days, but considering the age of the structure, it has weathered the years fairly well. The interior of the building is painted with pastel colors. Photos and artwork are hung on the walls throughout the building. Most of the artwork consists of abstract portraits and caricatures. With eyes peering at you from all directions, the paintings alone were enough to give you the feeling that you were being watched.

When our investigation commenced, it was a warm December evening with a temperature in the low sixties. The outside temperature was much warmer than inside the building because the brick walls held in the cold from the previous week's cold spell. Upon arriving, Kae gave us a tour of the premises. She told us the story of the Stillwell murder case and the connection that the house had to the murder. She also told the story of what the former tenant had seen and her reason for moving out. During the initial walkthrough, the place gave me a feeling that it was still occupied and I was invading someone's privacy.

Staircase at LaBinnah Bistro

We started upstairs with Kae leading the way and Gary just behind her. As Gary neared the top of the stairs, he stopped and gripped the handrail. He said that he felt a type of energy or vibration. In the paranormal field, Gary is what is called a *psychic empath*. This is a person who is especially sensitive to energy and its associated vibrations. All thoughts and feelings produce vibrating energy, and all of us release these into the collective surroundings. We all affect everything, even in complete isolation. The same holds true for our thoughts, words and intents. Everything created in solid space or through thought has an energetic charge that becomes available to all. A person who is an empath unconsciously deciphers others' energy and assimilates it as if it were innate. Gary picked up this vibration near the top of the stairs and asked if anyone had either died or been injured in that particular spot. Kae was not sure. However, with the steepness of the old staircase, and the tight turn at the top, it would not be out of the question that someone may have fallen and been injured at that spot, at some point in time.

We continued upstairs with the tour of the building and explored the area. At the top of the stairs was a room that I would describe as a sitting room. I placed my digital recorder on an end table and let it record the entire night, but to my disappointment nothing unusual was recorded. A small bedroom was located to the left, at the top of the staircase, through a doorway. There were several rooms upstairs, but the only room that I felt eerie or uncomfortable in was in this bedroom. Several times throughout the night, I had a feeling that someone was in the room with me when I was actually alone. It also gave me a feeling that I was being watched, but at other times I would go into this room alone and I would have a normal and comfortable feeling.

Once the walkthrough of the location was complete, we began unloading the equipment in order to get the investigation underway. Gary's equipment is quite extensive. It includes the standard equipment such as EMF meters, tape recorders, laser pointed thermometers, and an expensive digital camera. He also has more elaborate equipment, such as night vision goggles and Geiger counters.

Gary decided to set up a surveillance camera near the entryway of the dining area where the apparition was seen by the former tenant in hopes to capture video evidence. Upon setting up the surveillance camera, we hooked it up to a video recorder and TV monitor. We attempted to use the remote for the VCR to see if it was working. However, the batteries for the remote were dead. I gave Gary some new batteries for the remote and he placed the dead batteries in his vest pocket. The reason that I pointed out the dead battery issue is that it may have been significant to our investigation. One of the theories that paranormal investigators have is that spirits will drain energy from batteries, or other sources of energy, including people, in order to manifest themselves.

Later in the evening, when we were breaking down the equipment to leave, Gary took my batteries out and gave them back to me. He then put his old batteries in the VCR remote and, *lo and behold*, they worked just fine. Paul also experienced similar phenomena when using a handheld video camera that went from five hours of battery power to an almost a dead battery in a matter of moments, only to have the batteries recharge later just as inexplicably as they went dead. When doing a paranormal investigation, phenomena like this can be valuable in assessing whether a location may be haunted or not. Note to would be ghost hunters: bring lots of extra batteries.

The plan for the evening was for Gary to set up his surveillance camera in the dining room, Paul would walk from room to room using his handheld video camera shooting additional footage, and I would walk around taking still photos with my camera. Still downstairs, Kea's boyfriend went into the kitchen to look around. He flipped on the light switch and nothing happened. Moments later, I walked in behind him and flipped on the switch. The lights came on! Was the ghost of the house playing with us, or were we dealing with an outdated utility system?

Interior at LaBinnah Bistro

Once the equipment was set up, we did another walkthrough. Upstairs, in the bedroom that gave me the eerie feeling, Gary passed by a small loveseat and stopped. He sensed that someone was sitting there and that it was an older female who was in the house to look after a young child. He said that he sensed she was concerned that we were in the house. I took several photos of the area around the loveseat, but nothing turned up in the pictures. Later, I returned to this location several times by myself. On two of these occasions I had the strong feeling that someone was with me in the room. I guess you could say that it was a creepy feeling, although I did not feel any hostility. Other times, when I walked through this area, I would not sense anything at all.

Later that night while investigating the kitchen, Gary sensed that a child, possibly twelve to fourteen years old, was there with him. Kae had a strong feeling that the child was a boy and that his name was Nathaniel or Nate. I took several photos in the vicinity in which Gary said the child was standing, but nothing turned up in those photos either.

During our exploration, Gary and I discovered a door leading to the attic. I proceeded up the steps with Gary behind me with a light. I started to walk into the attic, but the floor was very weak and felt as though it would fall through if I placed my full weight on it. I snapped four or five photos and in one of the photos captured a very odd shaped orb which looks like it was in motion.

A short time later, Paul was using his video camera to record down the old staircase when he captured footage of a similarly shaped orb floating up the staircase. The orb moved as though it had a mind of its own. It even changed directions several times. I was also able to capture a photo of a similar orb at the foot of the staircase about an hour after Paul recorded the floating orb. It was as though it was following us.

About four hours into the investigation, we decided that we should start breaking down the equipment and call it a night.

I mentioned earlier that Gary's batteries for the VCR remote were dead and that this would be significant later. Well, as we were breaking down our equipment, Gary took the batteries that I had loaned him out of the remote and gave them back to me. When he did so, he put the old batteries back in the VCR remote. Just for the heck of it, he tried them and they were fully charged. So it was like something had *borrowed* the energy from his batteries. When they were finished using the energy, the batteries were charged again. Was this paranormal or just a malfunctioning remote? We may never know, but based on the location that we were at, it was very interesting *nonetheless*.

Investigation Summary

Anytime you get the opportunity to investigate a place with the history of the LaBinnah Bistro, in a town such as Hannibal, I would definitely consider it worthwhile. The investigation itself was a very good learning experience for how to conduct a proper investigation. I learned a great deal from Gary and Paul that night. Both are absolute professionals at what they do. They are more than willing to offer advice as well as to share their knowledge of the paranormal and filmmaking with those who have similar interests.

Is the LaBinnah Bistro haunted? Well, let us take a look at the evidence.

First, we have a secondhand, but credible, eyewitness account from the former tenant who described seeing a translucent figure moving across the dining area. She was frightened enough that she moved out. Furthermore, she had no reason to make up the story, so I would consider her account to be credible.

The draining of both Paul and Gary's batteries for no apparent reason also lent credence to possible spirit activity, although it could have been mere coincidence.

Paul's video footage of a floating orb, that seemed to purposefully move up the staircase, alongside my photos of similar looking orbs, both on the staircase and in the attic, offers more evidence of a possible haunting. Add to that, Gary picking up feelings that both a young boy and an older female were occupying the location. Specifically, Gary's feeling that there was an older lady in the same room in which I felt as though someone had been watching me.

Even with the combination of the above evidence and my personal experience of feeling like I was being watched, we still did not have enough conclusive evidence to declare the LaBinnah Bistro to be haunted. Outside of physically seeing an apparition, I feel that video, photographic, or audio documentation is necessary in order for me to determine whether a place to be haunted. Even though video footage of what seemed to be an orb moving with a kind of intelligence was caught on tape, it was still an orb. I was able to get several photos of orbs, but again, that does not prove that a place is haunted.

Many investigators believe that orbs are the lowest form of an apparition manifesting itself. I believe this to some extent, but I also believe that even if orbs are not simply dust or insects, they could be caused by other natural sources of energy or magnetic fields.

Unless evidence such as a photo or video of an actual apparition is captured, I cannot deem any place to be haunted. Secondary evidence such as feelings, cold spots, and photographs of orbs or mist should only be used to determine if a follow up investigation is warranted, but it is not enough to prove a haunting. With that being said, I believe that it is possible the LaBinnah Bistro is haunted, but we were not able to prove it during our investigation.

Kemper Military School

CHAPTER SIX

Kemper Military School

Booneville, Missouri

Located between St. Louis and Kansas City is one of the most interesting places that I have ever investigated, Kemper Military School in Booneville, Missouri. The institution, originally known as the Boonville Boarding School, was built in 1845 by Frederick T. Kemper and

CHASING SHADOWS

became a co-ed institute in 1862, when it was renamed Kemper Military School. At one time, Kemper was the oldest boy's school west of the Mississippi River, until it officially closed in 2002. Its most famous alumnus was Will Rogers, who attended the school in the 1890s. Rogers went on to gain worldwide fame as an actor, humorist, political commentator and performer until his untimely death in a plane crash in 1935.

Kemper Military School was the original choice for the location of several films, including *National Lampoon's Animal House* and *Taps*, but the school turned down both offers. It has been used as the setting for a number of other movies, including *Combat Academy,* a very low-quality takeoff on *Police Academy,* and *Child's Play III*. During the filming of *Child's Play III*, some cadets and instructors served as extras. In September and October 2007, Kemper's abandoned campus was used for location shots for the movie *Saving Grace,* which is about a little girl's trip back to Boonville in the summer of 1951, the year of the *great Missouri River flood*. Many downtown Boonville buildings were also used for filming, with Kemper the setting for an asylum. The movie, released in 2008, was directed by Connie Stevens and stared Penelope Ann Miller and Tatum O'Neal. The first investigation in which I was involved there took place in September of 2007 and we were not able to do a

139

follow up investigation until December of that year due to the filming of *Saving Grace*.

I became familiar with the school when Paul Robinson and Gary Hawkins asked me to help with the filming of a paranormal documentary about the haunting there. Over the years, many have reported seeing the apparition of a female cadet, either jogging or walking, near the track, but she seems to disappear when she nears the location where she was killed. According to accounts, the female cadet had broken up with her boyfriend, but he talked her into going for a walk with him. During the walk, he raped and killed her near a bridge located on campus, close to the jogging track.

D Barracks is reportedly the most haunted building on campus. Stories have been told about the fourth floor, in which a figure is seen standing in the window at night. The fourth floor was closed to students and when I asked a former cadet the reason why, he told me because no one could stay up there. He said there was something about the floor that made visitors feel uncomfortable and even scared.

People also claim to hear footsteps in C Barracks. Several years ago, while the barracks was being refurbished, construction workers reported hearing phantom footsteps. When they investigated, however, no footprints were found

CHASING SHADOWS

even though there was sawdust and dirt on the floor from the construction.

During our first paranormal investigation in September of 2007, the investigative team consisted of Gary, Paul, and I, along with a cameraman and sound man from the local Fox TV station from Columbia, Missouri. Paul was working for the FOX station at the time. He asked the pair to join us, to add additional camera and sound support. I do not remember the soundman's name, but the cameraman's name was Eric. I remember Eric because he was very skeptical of ghosts and hauntings when he arrived, but he had a little different outlook when he left. A writer from *Inside Columbia Magazine* named John and the hosts of an Internet-based paranormal radio show joined us as well.

At the time that we investigated Kemper, the property was owned by the City of Booneville, so in order to gain access to the buildings, a lady from the city met us and unlocked several of them including Barracks D and the administration building. She gave us a general tour of the campus and told us some of the stories that were connected to the hauntings. She also told us that a paranormal group from Kansas City, Missouri had attempted to investigate the building a few weeks before us, but when the group went into Barracks D, several of their members became sick to their stomachs and had to leave, so they called off their investigation. When our group entered the building there was the feeling of very strong vibe or energy present. We could feel the difference in the air. It was a heavy feeling that took our breath away. It was a feeling like you get after you have been running; a tightness in your chest. Immediately upon entering, three group members said, "Do you feel the heaviness in the air?" It was pretty uncanny to have three people bring up the same thing all at the same time. No one became ill, however, so at least we had passed the first test for the night.

After the walkthrough of the campus and its buildings, it was time to start setting up for the investigation. While the rest of the group went outside to gather up their equipment, I decided to take a few pictures inside of the kitchen in Barracks D. Everyone else had left the building and I was alone. I snapped several digital pictures and reviewed them, but there was not anything unusual. I then snapped another picture using the flash on my camera. About three or four seconds after my flash went off, another flash emanated from the back of the kitchen. It was like my flash had gone off again, but it had not. The flash came from across the room. I asked if anyone was there, and no one answered. So, again I said, "Is anyone there?" Again, there was no response. I walked to the back of the kitchen and no one was there. I then went outside and, sure enough, everyone was accounted for. I explained what had happened and they all assured me that they had been outside. I have no idea what caused the flash of light, as the kitchen is in the interior of the barracks so there were not any windows that would have allowed a flash to come from the outside. Plus, there wasn't any electricity turned on in the building, during the first investigation, so the flash did not come from any faulty lighting.

With one unusual incident under my belt already, I was more than ready to get the investigation underway. Once inside, the team did a sweep of the first floor together. Again, we ran into a spot in the building where the air felt very heavy and took our breaths away. While walking through, we started hearing what sounded like moaning or groaning, but could not determine where it was coming from. The sound guy was using a parabolic microphone and was wearing a headset. He too was hearing it loud and clear. Once we got near to what was the old mess hall, we started hearing a loud hollow banging sound, as if someone was banging on a drum or tom-tom! When we looked into the room that had housed the old mess hall, the banging stopped. We would hear the banging throughout most of the night, but could never find the source. Later that night, when a former cadet of the school unexpectedly showed up and we told him about the banging and other noises that we heard, he had an explanation for all. I will relate the cadet's stories a little later in the chapter.

After checking the mess hall, but not being able to determine where the banging was coming from, we decided to break up in small groups in order to cover the building better. The cameraman, Eric, and I would check out the basement area; while Gary, Paul, and the sound guy would cover the second floor. John, the reporter, and the radio host tagged along with Paul and Gary.

The basement was huge. Its layout was one long hallway or corridor going completely around the basement with many interior rooms branching off of the hallway. The lack of electricity in the building added to the creepiness of the basement. With every doorway, nook, and cranny we walked past, we got the feeling that something was lurking behind!

Eric and I walked through the basement hallways, filming and using the LCD screens of our video cameras to see our way around. As we were walking down one long corridor, I heard what sounded like either someone clearing his or her throat or groaning coming from behind me. I assumed that it was Eric. A few second later, Eric said, "Larry, did you hear that noise?"

"Do you mean what sounded like someone clearing their throat?" I asked

"Yeah," Eric replied.

"I thought that was you," I said. "But if whatever made the noise wasn't you and it wasn't me, then that means that whatever made the noise is between you and me Eric!"

I can't repeat what he said, but he was startled to say the least!

I have been to many dark cemeteries late at night, but I have never had a feeling like I had in the basement of Kemper. There was an overwhelming and constant feeling that someone or something was going to jump out at me. This feeling was there the entire time that I was in the basement that night. The groaning, or clearing of the throat, was the same sound that we had heard earlier on the first floor and that was picked up by the soundman's parabolic microphone. The odd thing was, none of us recorded the groaning sound on any of our equipment. However, when we returned to Kemper in December of that same year, we did not hear the moaning and groaning, but we recorded it with our digital recorders.

Later in the evening, Paul and I investigated the third and fourth floors. When the school was open, the fourth floor was locked and considered off limits. We were not sure why at the time, but we would find out later in the night when the former cadet unexpectedly arrived on the scene.

145

While on the third floor, Paul and I noticed how it was getting warmer in the building. There was not any activity on the third floor, so we decided to head up to the fourth. We tried the door on one end of the fourth floor stairwell and it was locked. Not only was the door locked, but it was double locked with a heavy, steel deadbolt. Paul and I thought that it was unusual to have such a heavy lock on a door to an empty section of the building. Since the door was locked, we made our way back down to the third floor and headed to the stairwell at the other end of the building. This time we were in luck, as the locks had been removed from this door.

When we entered the fourth floor, there did not seem to be anything special or different to distinguish it from the other floors. There were just empty rooms. One thing that I did notice was the temperature. The average temperature in the building was about 80 degrees and being on the top floor of the building, you would expect the temperature to be warmer than the other floors, but it was not. It was quite comfortable. I made the comment to Paul how I was surprised at how cool it was. Paul agreed, but he was not even finished with his sentence when I noticed sweat rolling off of his forehead and the temperature became noticeably warmer. One moment it was nice and cool and the next minute we were burning up! We continued to walk a few more feet and the temperature cooled down again. Rather than walking into a cold spot, we had walked into a hot spot. We could not find any reason why there was such a great fluctuation in the temperature.

Staircase to 4th floor at Kemper Military School

After we finished our investigation of the fourth floor, Paul and I headed back down to the second floor where we joined Gary, Eric, and the sound man. By that time, the reporter and the radio personality had left for the night.

We briefly explained to Gary what had just happened with the extreme temperature fluctuation on the fourth floor. While we were talking, all of a sudden the three of us heard a loud tapping noise, like *tap, tap, tap*. It sounded almost like someone was walking down the hallway. We searched the second floor, but could not find the source of the tapping. We heard the sound of the tapping several more times during the night. The former cadet also provided an explanation for that sound.

Since we could not find the source of the tapping, we decided to head back down to the basement, since Paul had not been in the basement yet. Gary decided to take a short break outside to get some fresh air, so the rest of our group headed downstairs without him.

We had been downstairs for a good twenty minutes when Paul's cell phone rang. It was Gary calling from the parking lot. He told us that a guy who claimed to be a former cadet had arrived and was telling him about some of the experiences that he had at the school. Gary thought that Paul might want to film an interview with him, which is exactly what Paul did.

The former cadet's name was Pierre. Pierre had attended the school in the 1970s through both Jr. High School and High School. He was very friendly and when he found out why we were at the school he really opened up. He asked us if we had experienced anything unusual. Paul explained how we had heard what sounded like moaning, tapping, and banging sounds in D Barracks.

Pierre told us the groaning was caused by the spirits of the victims of a double homicide and murder-suicide that had happened in the kitchen and the Mess Hall of D Barracks. It all started when a cadet was dating one of the kitchen helpers. Theirs was a rocky relationship. One day, the boyfriend came in and shot his girlfriend point blank, killing her. He then turned the gun on himself. Pierre believed their spirits continued to haunt the building.

As far as the tapping sounds, Pierre said that when he was a cadet, the officers would walk around the building making sure that lights were out and everyone was in bed. Only the officers were allowed to wear metal taps on their shoes, so at night after lights out, the cadets would hear the officers making their nightly rounds making sure everyone was in bed. When they walked around, he said he would hear the sound of the taps echoing in the hallway. The cadets did not want to hear the sound of the tapping stop in front of their door, because if they did, something was wrong and they were in trouble.

We asked Pierre if he knew what the hollow banging sound was and he said that yes, he knew that also. He explained that the officer of the day had a wooden desk that sat in the front of the Mess Hall. The cadets only had a short time to eat. At the end of each meal, the officer of the day read the daily announcements. To signal that it was time for the announcements to be read, he had a small metal rod that he would bang on the side of the wooden desk. Once the officer banged the metal rod against the desk, the cadets immediately stopped eating and laid all their utensils down on the table or else they would receive demerits.

When Pierre finished his story, he asked if we had been to the fourth floor. Paul and I explained that we had just been up there and we told him about our experience with the fluctuation in temperature. Pierre replied that this did not surprise him. He said that when he was a cadet, the fourth floor was off limits and was locked so that no one could access it. We asked him why it was off limits and he said that for some reason the floor gave off a bad vibe and that no one liked being up there, so the school closed it. He added that even though it was off limits and locked, if you were brave enough to go up there, the cadets knew a way to get in. Pierre told us about a friend of his that had a terrifying experience on the fourth floor. It seems that Pierre's friend was a tough guy and was a member of the school's football team. He did not fear anyone or anything.

Once, someone dared him to sneak up to the fourth floor and spend the night. So, he took the bet and snuck out of his dorm room one night. He went up to the fourth floor. Pierre said that his friend was not up there more than a couple of minutes when he came flying down the stairs like he had been thrown. He was trembling and nearly hysterical. Pierre asked him what happened and his friend said he did not want to talk about it. He said that he was never going back to the fourth floor again. Finally, Pierre was able to calm him down and his friend told him what happened. He was walking down the dark hall when he heard someone walking right behind him with what sounded like taps on their shoes. He then heard a voice that said, "What are you doing on my floor?" The friend was sure that it was one of the officers of the day, so he turned around and he told Pierre that what he saw was not an officer, it was not a human, and he was not sure what the hell it was, but he did not stick around to find out. Pierre said he had never seen his friend so scared before or after his experience on the fourth floor. He would never talk about the experience again.

A heavy, odd feeling to the place remained for most of the evening and into the night while we were investigating Kemper, especially in Barracks D, as though we were not welcome, but once Pierre arrived, the feeling changed. It felt like it was okay to be there as long as he was with us. As an ex-cadet, he belonged there, but we did not. We were intruders.

The only other interesting thing that happened that night was while we were investigating the old administrative building. Everyone was in the building, but Gary. Gary's back had started to give him problems, so he went outside and sat on a bench for a while to see if it would feel better. Nothing unusual was happening, when all of a sudden Gary hurried back inside. He said, "You will never believe what I just saw!" As Gary was sitting on the bench, he saw something or someone walking across the courtyard. After watching it for a few moments, he could tell that it was a person dressed in a cadet's uniform. Now, besides Pierre, no one else was on the campus that we knew of. The school closed in 2002. The cadet went behind a building and Gary never saw him again. I am not sure what Gary saw, but he saw something. He was very excited about it and he is not one to jump to conclusions. Gary is usually the first to try and find a logical explanation for anything he witnesses, but we all agreed that it would be highly unlikely that someone would be running around a closed military campus in a cadet's uniform at 3:00 a.m.

As daylight approached, we began to breakdown our equipment to get ready to make the four hour trip home. We had several personal experiences during the investigation, including the sound of tapping on the floors, banging, moaning and groaning, temperature changes, and Gary seeing the apparition.

In December 2007, I made my second trip back to Kemper. This trip was not as pleasant as the first, and as a matter of fact, it was quite treacherous. When I left my home in central Illinois, it was forty degrees and raining, but by the time that I met up with Gary in Alton, Illinois, the temperature had dropped to thirty-two degrees with sleet and freezing rain, and by the time Gary and I made the two hour trek from Alton to Booneville, Missouri, the temperature was down to twenty-three degrees and the freezing rain continued. At Kemper Military School, Gary and I met up with Paul Robinson and a couple of paranormal investigators from St. Louis, Missouri who specialized in high-tech audio. Our contact lady from the City of Booneville met us at the abandoned school and gave us a key to get in. She then left, but said that she might come back later to take part in the investigation.

The parking lot was like a frozen tundra. We had to find a convenience store to purchase a large bag of rock salt, to spread over the sidewalk and parking lot to melt the ice, just so that we could walk without falling down. Once we spread the salt and the footing was more stable, we began unloading our equipment. This investigation would entirely focus on Barracks D, as this was where most of the activity occurred during the September investigation. Even though Gary saw the apparition walking across the courtyard, the weather conditions were too extreme for our equipment to be set up outside.

It was going to be tough to detect any cold spots during this investigation, because at the time that we started, it was twenty-nine degrees inside the building. However, to us, just getting out of the wind made the temperature feel warmer.

Paul and the two-team members from St. Louis decided to set up our base for the investigation in the old mess hall of the barracks. While they were doing this, Gary and I started setting up surveillance cameras in the kitchen and a stairwell that led up to the second floor. The kitchen was in a separate room from the mess hall, and where we were, it was probably seventy-five to one hundred feet from where the other team members were setting up base camp.

After about thirty minutes, as Gary and I were finishing setting up our cameras in the kitchen, we heard a woman's voice coming from the mess hall. She was talking loudly and laughing. I said to Gary, "Who's the woman that Paul and the guys are talking to? Did the lady from the city come back already?"

"I don't know," Gary replied, "but I wonder what's so funny."

Then I said, jokingly, "I don't know, Gary, but it's just like Paul to find a girl and hold out on us so he can have all the fun!"

Gary replied, "Let's finish setting up this last camera then go see who is here. We need to tell them to tone it down so our video cameras don't pick up their conversations."

Gary and I finished setting up the cameras and then headed back to the mess hall to see who the lady was. When we walked into the hall, Paul and the two investigators were setting up equipment on a table, but no one else was in the room. I asked Paul, "Paul, did the lady from the city come back?"

"No," he replied, "why?"

"Well," Gary demanded, "who was the girl that you guys were talking to and what was so funny?"

"There's been no one else here, but the three of us," Paul replied. "Why do you think there was a girl here?"

We explained that both Gary and I had clearly heard a conversation and the laughter of a woman that seemed to be coming from the mess hall. Her voice was loud and she seemed to be having fun. Paul again assured us that no one else was there, but the five of us and that they had not heard the voice or laughter of a woman. Unfortunately, at the time Gary and I heard the voice, our equipment was not recording yet because we were still setting up. Due to this occurrence, and so as to not miss out on possible evidence again, I now turn on a recorder, carrying it with me as soon as I enter a location.

Gary and I began to wonder if we had heard the voice of the female who was murdered in the kitchen at the hands of her former boyfriend. I mean, who else would haunt this area of the building and why? This experience was definitely one of the weirdest experiences that I have had as a paranormal investigator. It would have been one thing if I was the only one to hear the female voice, but it is quite another when you have a witness who heard the same thing!

Unfortunately, this would be the only personal experience that we would have that night. The rest of the night turned out to be quite mundane.

When I returned home and reviewed the evidence from our investigation, I discovered that I had recorded some groaning sounds. It was the same groaning that we had heard all night during the September investigation, but did not hear it at all during the December investigation. This was very odd, and I do not have an explanation for the discrepancy. It is just another mystery of the supernatural.

Investigation Summary

Aside from the Villisca Axe Murder House, Kemper Military School is the only other location that I have physically heard unexplained voices. Add to that, the moaning sounds that we heard during the September investigation and recorded during the December investigation. The female voice and laughter that Gary and I heard was also extraordinary. The voice was coming from within the building and sounded like it was coming from the very next room. Additionally, Gary saw an apparition of a cadet in the courtyard, and the entire team heard tapping sounds on the tile floors of the school, as well as hollow banging in the mess hall. These were things that we experienced. Pierre, the former cadet, gave us plausible explanations for what we were hearing, but if he was right, then they are of supernatural origin. I always try to find a natural or logical explanation for things that I experience, but there is not a logical explanation for voices and sounds coming from thin air!

All the time that I was on the grounds of Kemper, I had the feeling that activities from bygone days were continuing. I am not sure whether they are residual or possibly some type of time loop, which is somehow able to bleed through the veil from the other side and into our physical space. It is experiences like these that have lead me to believe in other realms or dimensions. If what we have been taught to believe about the existence of Heaven and Hell is true, then other realms or dimensions do exist. If these other realms and dimensions co-exist, then so does the possibility for interaction. EVPs, along with the occasional photographic and video footage, are evidence that this interaction is taking place. The question that I am trying to answer as a paranormal investigator is "How does this interaction take place?" Once we answer this question, a whole new world of possibilities will open up for mankind.

Morse Mill Hotel

CHAPTER SEVEN

Morse Mill Hotel

Morse Mill, Missouri

The Morse Mill Hotel, or the *Blue Lady*, as current owner Patrick Sheehan calls it, is located just west of St. Louis near the small town of Hillsboro, Missouri. Originally built in 1816 as a one-room house, it was later expanded to its current size of 5,300 square feet, but there is some confusion as to when this expansion took place. Some say John Morse expanded it in the 1830s, and others say Morse expanded the building in 1847. I am not convinced of the accuracy of either source. Patrick Sheehan, who specializes in restoring old buildings, recently purchased the hotel.

John Morse was a premier bridge builder and engineer who used the Morse Mill as his family residence. Since then, it has been used as a hospital for Confederate prisoners of war, a hotel, a brothel, a speakeasy, a United States post office, and a half-way house. It is also believed to have been a stop on the Underground Railroad.

The Morse Mill was also home to the first known female serial killer in America, Bertha Gifford. Gifford was born Bertha Alice Williams in 1876 and died in 1952, in a mental institution. She lived in Missouri where she acquired a reputation as an angel of mercy by showing up at the homes of sick friends and family, many of whom subsequently died. So many died that their deaths began to make people suspicious, and Gifford was eventually arrested for murder. She was tried and found not guilty due to an insanity plea, and she spent the rest of her life in a mental hospital. Gifford was believed to have murdered anywhere from seventeen to twenty-three people and possibly more. A headstone with the name Bertha Gifford is located in a small cemetery down the road from the Morse Mill. According to some, however, she is not actually buried there.

Many other famous people are known to have frequented the Morse Mill Hotel. Frank Dalton, who claimed to be Jesse James, Al Capone, Charles Lindbergh, Charlie Chaplin and Clara Bow have all walked its hallways. It is located in Jefferson County, which was settled in 1799 by Francis Wideman, who was believed by many to be a sorcerer According to his own brother, pJohn, Wideman could "conjure up the devil."

John Morse came to Jefferson County in 1847 and built a commercial gristmill on the Big River. He used slave labor to quarry the stones for the building. He also built a home in the town, which was named after his first commercial venture, Morse Mill. Morse would go on to be a state politician, own at least two general stores, a contracting company, and a hotel. His home was used as a stagecoach stop and, after his death, it became the Riverside Hotel. Eighteen sleeping rooms were added onto the home to accommodate all the guests. Morse Mill became a resort town, with people coming to enjoy the slow, cool waters of the Big River.

I first heard of the Morse Mill Hotel while doing an Internet search for haunted locations to investigate. I found an article about a documentary called the *Morse Mill Project* that was filmed in November of 2008. The article described how, during the filming, the investigators had reported seeing the shadow of a large man. They told of cameras levitating off the floor while turning 360 degrees. They also described how, while they were upstairs, they heard a loud metallic sound, and when they returned downstairs to investigate the noise, they found an iron fireplace poker twisted in a U-shape. One of the filmmakers had been scratched through her clothes by some unseen force. The article also described how one group of first-time ghost hunters were slapped by unseen hands and had air blown in their faces. They also heard footsteps.

According to the article, during the first walkthrough of the building, one investigator was scratched down the neck by what appears to have been three unseen fingernails. I found another article in *Haunted Times Magazine* in which, during the time the Morse Mill was a homeless shelter, the residents there saw a black shadow person, which was described as some nine feet tall. Faces were even seen in mirrors. Overnight guests have told stories of shadow people walking about the property and down the road that passes by the hotel. These stories sounded almost too good to be true, so I had to see for myself.

I emailed a psychic from Iowa who was listed in the article as one of the investigators involved with the filming of the *Morse Mill Project*. In my email, I asked her if she could provide a phone number for Patrick Sheehan, the owner of the hotel, so that I could contact him about doing my own investigation. She agreed, and I called immediately.

When I explained to Patrick that I would like to investigate the hotel, he was very receptive. His only real question to me was how much experience I had as a paranormal investigator. I explained to him the number of years that I had been investigating, and also related some of the places that I had investigated. After I provided this information, he agreed to allow me access to the hotel. He explained some of the things his girlfriend and he had experienced, like the time they were standing outside the hotel, near the deck. He gave his girlfriend a hug, when suddenly it sounded like every window in the hotel rattled and shook. Another time, Patrick explained how a group had come to investigate late in the afternoon and planned to camp out, so he led them to the location on the property where they could set their tents up. His girlfriend was with him. She later told him that, as Patrick and the group were walking, she observed some type of a black mist or fog following them for a while, before ultimately disappearing.

Patrick's girlfriend also witnessed a lady standing on a balcony on the attic level of the hotel. She said that the lady was there one minute and then seemed to suddenly fold up and disappear!

The most fascinating story I have heard about the hotel was told to me on June 10th, 2010. Patrick asked me if I would come down to be interviewed for the TV show *Most Terrifying Places in America*, due to some of the EVPs that I had collected on my three investigations there. Unfortunately, the film crew never interviewed me due to lack of time. However, while I was there, I was introduced to a man named Jeff. He told me of an incident that happened to him and two other investigators when they spent the night at the hotel, during an electrical storm.

According to Jeff, his group set up several flashlights on the staircase in the hotel, and were asking the spirits to make the flashlights light up, to demonstrate that they were there. Jeff said that this seemed to be working right on cue. As they were doing this, a furious storm came up and a tornado actually touched down about two blocks away. Jeff and another investigator were facing the staircase and were monitoring the flashlights when a bolt of lightning lit up the staircase. When the lightning lit up the staircase, they could not believe their eyes, because they saw a little girl peering at them through the railing. When the flash of lightning disappeared, so did the little girl. They never saw her again. Jeff seemed very sincere in his story and I believe that he saw something.

I had my own strange experience there prior to my first investigation of the hotel. Before I investigate any place, I usually try to arrange a walkthrough days or weeks ahead of the actual investigation in order to get a feel for the layout of the building. So, I arranged to meet with Patrick, the owner, several weeks before. The afternoon I arrived at the hotel was your basic hot and muggy day in July. Since it was so hot that day, I wore khaki shorts with cargo pockets. When I arrived, I parked my SUV to the side of the building, put my keys in one of the cargo pockets, and left the vehicle unlocked. I met Patrick and we exchanged greetings. He then proceeded to give me a tour of the hotel and surrounding property. After the tour, I returned to my vehicle to retrieve the magazine article on the Morse Mill Hotel that was in the *Haunted Times*, because Patrick had not seen it. I opened the door and gathered up the magazine. As I did this, my eyes glanced at the vehicle's ignition, which caused me to automatically check my pocket for my keys. When I checked my pocket, I panicked because I could not feel them. I stepped out of the vehicle, checked all of my pockets, and even turned them inside out. My keys were gone!

I returned to the owner and handed him the magazine and said, "Patrick, I have lost my car keys." I explained how I needed to find them because my wife would not be a real happy camper if she had to drive three hours to the middle of nowhere to bring an extra set of keys. Patrick and I went back inside the hotel and retraced our footsteps, starting in the basement and ending up in the attic. The keys were nowhere to be found. We headed outside and I knew that finding the keys in the tall grass that surrounded the hotel would be nearly impossible. Just as we began to look around outside, my right hand brushed the side cargo pocket on my right leg. When I did this, my hand hit something and it rattled. When I put my hand in my pocket, I immediately felt my car keys!

Now, that had been one of the pockets I turned inside out earlier and the keys were not there. There was no way that I could have missed the keys when I turned my pocket inside out. "You are not going to believe this, but my keys are in my pocket!" I yelled at Patrick, who was in some tall grass looking for my keys. I explained how there was no way that I could have missed them, since I had both placed my hand to the bottom of the pocket and had turned it inside out. The keys simply were not there. Patrick began to laugh and explained how that seemed to happen all the time at the hotel. He explained how his tools have disappeared even though they were right next to him while he was working in the building. He would search for the tools and would later find them right where he had left them. The same thing happened with his keys on several occasions. They would mysteriously disappear only to reappear just as mysteriously.

CHASING SHADOWS

Other strange things have occurred at Morse Mill. One afternoon, soon after Patrick had purchased the hotel, he was working downstairs on the first floor. Patrick is a contractor and inventor of a commercial air cleaning system by trade, so he does most of the restoration work himself. He explained how he stored some of his tools in a locked room on the second floor. Not only were they locked up, but also he had personally installed heavy steel locks and latches that were bolted in the wood doorframes by heavy gage steel screws. That particular day, he was alone in the building. Suddenly, he heard a loud banging and screeching upstairs as if metal was being torn apart. He rushed upstairs and found the door to the tool room opened. He told me that the room was locked when he arrived and he had not unlocked the door. Not only was the door opened, but also the heavy gauge steel latch had been pulled off the door and was lying in the center of the hall with the lock still locked and attached to the latch. He had no explanation for this, but did not waste any time in re-bolting the latch and lock on the door to once again secure the tool room. When he left for the afternoon, he had a feeling that something was there, something that he could not see

Patrick also told me that, while working alone in the hotel, he has occasionally had the overwhelming feeling that a very large man was standing behind him. This *man* seems to follow him around and watches what he is doing.

Patrick also showed me the fireplace poker that had been mysteriously bent in a U-shape during the November 2008 filming of the documentary *The Morse Mill Project*. If something supernatural bent the fireplace poker, it would have taken a tremendous amount of energy and strength. One of my theories about ghosts is that they are from the quantum realm and are made up of virtual particles. Virtual particles borrow energy to exist, and many paranormal researchers believe that ghosts also borrow energy to physically manifest themselves. This may be one of the reasons that batteries in cameras, tape recorders, and other equipment used in paranormal investigations, seem to drain without explanation. Then, just as inexplicably, recharge again when they are brought outside the affected area. The Morse Mill Hotel does not have electricity, so can you imagine the amount of energy that a spirit would need to borrow in order to bend a quarter inch thick steel fireplace poker. If the story was true, where did the energy come from to perform such an incredible feat?

With the experience that I had with my disappearing car keys, and from the stories that Patrick had told me as well as the other stories that I had read about, I was totally psyched to investigate the Morse Mill Hotel.

The paranormal team that I put together consisted of Jamie Sullivan, Gary Hawkins, Dave McCracken, a member of Gary's team, and myself. We met Patrick at about 5:00 p.m. He gave us the key to the building and basically said to have at it. Patrick does not really like to spend much time at the hotel during the evening and rarely if ever does he like to be there after the sun goes down. So, as soon as Patrick left, we began to unload our equipment. After we finished, we conducted a walkthrough of the building. We started with the basement and worked our way up to the attic. The basement was very interesting, to say the least. The foundation and walls of the basement are made up of limestone. Any paranormal investigator will tell you that if there is paranormal activity at a location, limestone will generally be found on the property. This does not mean, as some skeptics believe, that all locations having limestone should be active, as they are not.

My contention is that, since limestone constitutes approximately ten percent of the sedimentary rock exposed on the earth's surface, the availability of it as a building material is why many older structures were constructed using limestone blocks as walls or foundations. Skeptics are correct, in that, not all buildings built with limestone, or properties that have large quantities of it lying around, are haunted or have paranormal activity. But by the same token, not all cemeteries seem to be haunted either. In my experience, especially with older buildings and even cemeteries, fairly large quantities of limestone are present at places that are active. It also seems as though most of the locations where paranormal activity takes place have had some kind of incident, be it by tragedy or just a memorable event, that have left a burst of energy or higher vibrational imprint behind, because of the emotions caused by the event. That is only my theory, but there is not a skeptic on the face of the earth who can scientifically prove that I am wrong. By comparing anomalous events and personal experiences, with data from my investigations, I hope to someday be able to prove or disprove my own theories.

The basement is divided up into several rooms. In one room, there is a large stone fireplace. This is the room that was used as the speakeasy. To the rear of this room is an old cistern. Patrick, the owner, who is a well-respected commercial contractor, told me how unusual the cistern is. Even its size is unusual; the opening is about twelve feet across. When they started cleaning out the cistern, they found all types of debris in it. Patrick said that the walls of the cistern were made of fourteen-inch handmade limestone blocks, but once they made their way down about fifteen feet, the walls became only a few inches thick. Structurally, this did not make sense. Patrick thought that there may have been a tunnel entrance there for the Underground Railroad, so they chiseled their way through the stone wall of the cistern, but no tunnel was found. The last time that I talked to Patrick, they had made their way down to about eighteen feet. They were still finding debris and had yet to reach the bottom.

At one end of the basement there is a small dark room with a large limestone block in it. You can see holes in the block where bolts were once secured. The bolts were used to shackle slaves to keep them from running off. Although abolitionists brought runaway slaves there to teach them how to read and write as they were being moved further north, some of the slaves did not understand English, so they would try to escape if they were not shackled, which could have led to their recapture and return to their former masters.

Once we finished with our walkthrough of the basement, we worked our way up to the first floor, second floor, and then the attic. The only encounter that we had was in the attic, but it was of a more physical kind. There were hundreds of wasps that had entered the attic through several of the broken windows, so in order to investigate; we had to take a short road trip to purchase several cans of wasp spray. Once we took care of the wasps and sprayed the broken windows, the problem was solved.

There were no spikes or fluctuations in the temperature or the electromagnetic field, and the attic did not have any paranormal feeling to it at all. It just seemed like your normal, everyday old building. The entire night, from 5:00 p.m. the evening before and until the next morning at 5:00 a.m., was uneventful. However, when we reviewed the audio, we did record several interesting EVPs.

We recorded several EVPs during the investigation, including ones that said, "Mary Hudson," "Quiet Please," "And I would watch you," and "Where are you, Henry?" The best was the sound of someone strumming a chord on a guitar, then singing, "I love you." It sounded like something that you would have heard Jimi Hendrix singing in the 1960's. Despite the EVPs, I was disappointed that we did not have any personal experiences.

One year later, I returned to the Morse Mill Hotel with Craig Whitworth, a paranormal investigator from central Illinois. The hotel seemed to be more active that night. Once again, we had no personal experiences, but we did get high EMF readings on the second floor and there seemed to be a pattern to the activity. After we started getting hits on our K-II Meter, we noticed that if we stood by the doorway of a particular room, we would get a spike on the meter. We noticed that that this spike would occur, then dissipate and would later return. The hotel does not have any electricity, and none of our equipment was causing the spike in EMF readings. We left our cell phones down in the kitchen. We also noticed that the EMF readings seemed to move from room to room, and it got to the point where we could actually follow the EMF readings around the second floor. It is possible that this may have been some type of a residual activity, but the energy source was what puzzled us. Our equipment had not experienced any unusual battery drainage. The energy seemed to be coming from an invisible source.

Later in the evening, we had another unusual occurrence in a room on the second floor. The building's owner had told me a story about this particular room during my walkthrough of the hotel, the previous year. According to Patrick, people who stay overnight in this room seem to be awakened in the middle of the night feeling like they have been touched in inappropriate places. On one particular night, a pair of paranormal investigators, a man and woman, decided to spend the night. After investigating for several hours, they decided to call it a night and go to sleep on a mattress that was in the room. In addition to the mattress, there was an old electric fan. The fan was not plugged in, since the hotel does not have electricity. During the night, the male investigator awoke to the feeling of being touched in a very personal place. Startled, he arose to see who was there. When he looked over at the fan, he could see the electrical cord floating in the air. This startled the couple so much that they left in the middle of the night.

On the night of our investigation, Patrick had shown Craig and I, a small section of the floor that had been cut out by the previous residents of a halfway house, which used to be located in the building. It was like a secret compartment in the wooden floor. It had a hole in it so that a person could put a finger in the hole to lift up that section of the floor. The section was about four by twelve inches and had been used by the residents to hide drugs. A bed sat over it to conceal its existence.

During the night of our investigation, Craig and I started doing an EVP session in the room by asking questions and using the K-II meter as a control device for obtaining answers. Before we started the EVP session, we said out loud that if any spirits were present, they could communicate with us by making the lights on the K-II meter light up. The response would be two flashes for yes and three for no. We began asking questions and it appeared we were getting responses soon after. We started with a control question. The control question that I always use is, "Is Fred Flintstone present in the room?" Of course, if the light responds with two flashes for "yes," we know that it is only an electrical anomaly and is not paranormal. Or, at the very least, we know it is not an intelligent haunting. It is still possible that the K-II meter is picking up energy from a residual haunting.

At one point during the EVP session, Craig asked, "If I find something in the compartment, can I keep it?" Immediately, the light on the K-II meter began flashing so fast you would have thought we had just placed it next to an electrical circuitry box. When I played back the video of the incident, a male voice saying, "Go Shit!" can be heard. I guess the answer was no!

Another anomalous event happened as we came downstairs. I came down first and Craig followed. He was using a helmet camera he had designed using a hardhat. The camera was mounted on the hardhat facing backward so that he could film what was behind him. At the same time, he carried a handheld camera to record what was in front of him. This special camera setup also records audio. Craig is very innovative in gadgetry and he has worked in university research in the field of audio, so he is also very good at debunking false EVPs.

We had another camera setup in the doorway of the parlor, filming toward the bottom of the staircase. As Craig reached the bottom of the staircase, he mentioned that he was getting some type of interference on the handheld camera screen, which amounted to a rolling picture with lines of interference. I noticed that Craig turned and glanced behind him. Later, he told me that he thought someone might have been there. So, basically, Craig looked back at the same time that he got interference on his video screen. When we replayed the video footage from the camera in the parlor doorway, sure enough, we could hear and see Craig turn halfway around and glance back up the stairs. To our surprise, the video camera also recorded the voice of what sounded like a man with a strong east coast accent, or what sounded like a gangster from the 1920s, a time when Al Capone was known to frequent the Morse Mill Hotel. The voice simply said, "Yo Let Me," as if to be saying to Craig, "let me use the camera." What was even more interesting was that you could time Craig turning around and looking behind him to the voice saying, "Yo Let Me." The voice was recorded on Craig's video camera as well as the video camera in the doorway of the parlor. We also recorded an EVP that says either "Missed your Angels" or "Mr. Angels," but I am not sure what that means.

Investigation Summary

The rich history of the Morse Mill Hotel makes it worthwhile to investigate. It is not every day that you visit a place that was a hospital during the Civil War, part of the Underground Railroad, a Speakeasy during the Roaring '20s, frequented by Al Capone and his men, a hotel managed by the country's first female serial killer, and a place that had guests such as Frank Dalton, Clara Bow, Charlie Chaplin, and Charles Lindberg. Just stepping foot inside such a place gives you a rush of energy.

Other than the high EMF readings, I did not have any personal experiences of the paranormal kind. However, you cannot ignore the EVPs, and what seemed like responses to questions that Craig and I had asked, which registered on the display of the K-II meter. You also cannot ignore Craig Whitworth's simultaneous experience of interference on his video screen before turning around because he thought that someone was behind him. At the same time, he recorded a voice on two separate devices, which said, "Yo let me!"

If you asked me if there seems to be something unusual going on at the Morse Mill Hotel, my answer would be yes. At times there seems to be some kind of intelligence at work, which was apparent in several of the EVPs that I obtained. But I did not see anything as extreme as what was reported by the crew that investigated and filmed the *Morse Mill Project*. I do believe the story that Jeff Green told me about seeing the little girl in the flash of lightening, because Jeff was very passionate about what he saw. He certainly saw something. When you add up all of the above, they do point to something paranormal going on. Our definition of the paranormal is *something that is beyond the normal*. What goes on at the Morse Mill Hotel seems far from normal. It is definitely a place to put on your paranormal to do list.

Villisca Axe Murder House

CHAPTER EIGHT

The Villisca Axe Murder House

Villisca, Iowa

Over the years, most of us have heard at least one ghost story. As children, we believed these stories to be true. But as time passed and we grew into adulthood, most of the stories were forgotten, forgotten that is, until we had our own encounter with the supernatural. The story you are about to read is true and a matter of personal record. I cannot explain it, nor can I prove it. But to me, no proof is necessary, because I lived it and, as we all know, *seeing is believing*!

In my eight years as a paranormal investigator I have investigated many types of places, from cemeteries, to old mansions, military schools and private residences. I have experienced some unusual phenomena and unexplained events. However, nothing compares to the experiences that I have had both at and after investigating 508 East 2nd Street in Villisca, Iowa.

Villisca, Iowa is a small town located sixty miles from Omaha, Nebraska. According to the 2000 U.S. Census, Villisca has 636 housing units, but none are more famous than the one located at Lot 310, now known as the *Villisca Axe Murder House*. For those not familiar with the history behind the Villisca house, here is a brief summary of its incredible and horrible story.

On Sunday evening, June 9th, 1912, Joe Moore and his wife Sara took their four children, Herman, Katherine, Boyd, and Paul, to the Children's Day service at the Presbyterian Church. Accompanying them were Ina and Lena Stillinger, who had asked their parent's permission to stay overnight with the Moore children. The Moore family and their two guests, the Stillinger girls, returned to the Moore residence that night never to be seen again.

CHASING SHADOWS

During the night of June 9th, an unknown individual entered the Moore home and bludgeoned to death the eight people sleeping there. On June 10th, when residents discovered the eight victims inside, the small town of Villisca was changed forever.

Since the time of the murders, the house has had several families live there, but in 1994, it was in danger of being torn down. Local residents Darwin and Martha Linn, owners of the local Olson-Linn Museum, decided to buy the historic home rather than see it demolished. Darwin and Martha were instrumental in having the house put on the national historical register. They believed that the mystery of the crime would attract tourists to Villisca and its *Axe Murder House*. Little did they realize that its reputation for the paranormal would attract more interest than the infamous murder!

The new owners restored the house to its original appearance as it was in 1912. Darwin Linn replaced electric lights with oil lamps. Using old court documents and newspaper articles, the Linns decorated the rooms as they might have looked at the time of the murders.

Many local residents are not happy with the Linns, claiming that Darwin and Martha remodeled the house in order to capitalize on its gruesome history by allowing ghost hunters to lease the home to conduct investigations. I know, however, that this was not the case. The Linn's intention was simply to preserve an important part of the history of Villisca. Like it or not, and as gruesome as the murders were, they are a part of the town's legacy.

Darwin explained to me that he had never heard the term paranormal until after the remodeling on the house was completed and a man asking how much it would cost to conduct a paranormal investigation there contacted him. Once the man explained what a paranormal investigation was, Darwin saw no harm in it and agreed. So as they say, the rest is history. Due to the interest in the house by paranormal investigators, the Linns allow groups and individuals to conduct overnight investigations there for a reasonable fee.

This is where my story begins.

It is a story of how one house has changed the way that I approach investigating the paranormal. It is also a warning to those like me who are interested in delving into the unexplained and the supernatural, because we may not be fully prepared to handle the things that we find or come into contact with. The Villisca House has made me realize that paranormal investigation is not simply fun and games or a field that is not without its risk. Paranormal investigation is a field that we need to approach with caution and care. We need to prepare ourselves and our team members to take precautions, because once the door opens to the other side, what we find can be a living nightmare, a nightmare that can take its toll both mentally and physically.

I have investigated the Villisca House on four separate occasions, with my first visit in June of 2008 and the last on October 10th, 2010. It was during my second and third visits to the location, which had a profound impact on me as a paranormal investigator.

During my second investigation, in September 2008, my outlook on the paranormal, and life in general, was dramatically affected.

Paul Robinson, a paranormal investigator and film producer from Columbia, Missouri, Sharon Wright, a model/actress from Kansas City, Missouri, and I traveled to Villisca to film a documentary on the Villisca Axe Murder House. The owner, Darwin Linn, agreed to allow us to film there on September 8th and 9th free of charge. This was Paul and Sharon's first trip to the Villisca House. Paul is an experienced investigator, but this was Sharon's first experience with the paranormal.

With a couple of exceptions, our personal paranormal experiences were very limited during the investigation of the night of September 8th. At around 10:30 p.m., I was in the parlor and walked past a door that leads to the front porch of the house. The door was locked from the inside and there was also a screen door that was locked with a padlock on the outside, so both doors were locked and secured. As I walked past the door to look at a picture of the family on the wall, the door violently rattled and shook. It was as if someone took their foot and tried to kick it open. I was a good eighteen inches away from the door, with Sharon following close behind. Needless to say, it startled both of us. The floor was old and wooden, so in an attempt to recreate the violent shaking of the door, I stomped as hard as I could, but nothing happened. The door did not budge. I also pulled hard on the door handle and I could barely make the door rattle. Whatever caused the door to violently shake could not be recreated.

Blue room at Villisca

The next unusual thing that happened took place while we were in the downstairs bedroom, which was called the blue room. It was called the *blue room* simply because it was painted blue. This was the room where Ina and Lena Stillinger were murdered. All three of us were in the room. I was using brass rods and checking to see if there were any spots of EMF activity. As I walked into the bedroom with the rods, I followed the wall to my left and walked to the far corner of the room. At that point, the rod in my left hand started to spin in a counter clockwise direction. The rod in my right hand then started to spin in a clockwise direction. The rods were spinning so fast that I had to grip them tighter so as not to lose control of them! While this was going on, I had the sensation of an electrical charge shooting up both of my legs. As mentioned earlier, Darwin Linn removed all of the utilities in the home to put it back to the original state, so there was not any electricity in the house. Paul also held the rods and they reacted for him as well. Sharon was too spooked by their unexplained spinning to do the same!

I later found out from Darwin, that the spot that Paul and I stood in, was the exact spot where the axe had been left after the murders.

We stayed the night, completed our investigation, and then left the house at about 4:30 a.m. We returned the following day to finish filming the documentary, and since we had all the footage needed to complete our project, we did not stay the night of September 9th.

It was not until I returned home and reviewed my audio that I realized all the activity that had been going on right under our noses. I recorded no less than twelve EVPs, of which a few were somewhat disturbing. In one EVP, a cynical laugh and heavy breathing can be heard. In another, I could hear what can best be described as growling, but growling in syllables. By using audio enhancing software called Audacity, I was able to enhance the EVP by altering the tone, pitch, and speed of the audio clip. The *growling* turned out to be two different voices. First, the voice of what sounds like a child saying, "Help Me!" can be heard, then a second voice also says, "Help Me," but in a mocking tone. Immediately after the second "Help Me," you hear a voice clearly say, "Satan Wins!"

The frightening thing about this is that I had recorded this clip soon after we arrived at the house. Paul and Sharon were going to film some of the narrative voiceover outside. Owner Darwin Linn was outside talking to us, when I told the group that I was going to head on inside to look around, as well set up a couple of digital recorders and let them record for a couple of hours, while no one was in the house. Soon after I went inside, I went upstairs and was in the room where the four Moore children were murdered. As I was looking around, I heard someone walking around downstairs. I could hear Paul and Sharon's voices outside, so I knew that they were not in the house. I figured that it must be the owner that was inside, so I yelled downstairs, "Hey, Darwin is that you?" No answer! I yelled again, "Hey, Darwin," and again, there was no answer.

I walked across the upstairs hallway and looked out the window that faces the barn. As I looked out the window, I saw Darwin already in his truck, backing out of the driveway, so Darwin had not been inside. I went downstairs to look around and saw that no one was there. I was alone in the house. I went back upstairs to place my audio recorder in the children's room. I always announce the date, time, and the location of where I am recording at the beginning of the session. When I played the audio back from this particular recording session, I had recorded the EVPs saying, "Help me," "Help Me," and "Satan Wins," just as I had turned on the recorder, because the words appeared prior to my announcing the date, time, and location of the recording. Later, I often wondered if the footsteps and the growling voice belonged to whatever followed me home! Two of the four times that I have investigated the house, I spent a short time alone in the house. Both of those times, something followed me home! Is being alone in the house the reason why something chose me to follow home and torment, or is it merely coincidence. If you continue to read my books, you will soon realize that I do not believe in coincidence.

The most disturbing EVP was one in which my name was clearly whispered. In that particular clip, a creepy male voice first says, "Come here," immediately followed by a whisper that says, "Larry." Seconds later, a different male voice says, "I'll tie her," and then, in a matter of fact way, a final voice says, "I'll kill her."

In the four investigations I have conducted at Villisca, I have recorded more than thirty EVPs, most of which are very clear. The EVPs themselves were very exciting, not to mention interesting from an investigative point of view, but once I returned home from Villisca, things started to get very interesting indeed. I truly believe that something sinister followed me home on two separate occasions. It took a while for me to realize that something was not right, as too many negative things were happening around me to get a sense of it all at once.

The first thing I noticed was that a lot of electronic devices malfunctioned. In one incident, my son and a friend of his, were commuting to a local junior college, both of them attend, about thirty miles from our home. My son and his friend would take turns driving. On the Monday after I returned from Villisca, I received a call from my son at college saying that his friend's truck would not start. I went over to the college to see if I could start it for them, and the battery was dead. I tried to jump start it without success. Later that evening, the father of my son's friend jump-started the battery without any problem. The very next day, my son called from school again. He said, "Dad, you won't believe this, but my truck won't start." So I went over to the college and jump-started his truck. This time, it started. As a precaution, I left my vehicle with my son and I drove his vehicle to a relative who owns an auto mechanic shop. I had him test the battery and check out the truck. The battery was good and everything was working properly.

The next day, our refrigerator stopped working and all the food in the freezer spoiled. The next weekend, I was turning fifty and my wife had organized a birthday party for me at a new restaurant that a friend of ours owns. He has a party room equipped with three big screen TVs and a nice sound system, and my wife had made several DVDs that they were going to show. Mike, my friend and owner of the restaurant, put one of the DVDs into the DVD player to test it, and the system would not work. He told me that had never happened before. He fiddled with the system for over an hour before it just started working again without any explanation. He has never had this problem since. Then there were other, minor problems. Things broke on our automobiles, our computer crashed, and we got flat tires.

Our washing machine broke down, then, after visiting a friend's house, their washing machine broke down as well and so did their dishwasher. We also had trouble with our phone line, and so did our friend.

After I returned from Villisca, I had a strong feeling of being depressed. I have never had a problem with depression, but this was an overwhelming feeling that was made worse by irritability. It seemed like any little issue became a major issue. I began to notice that people who were around me were also becoming increasingly irritable and would get angry at each other over little insignificant things. I also started to notice that this seemed to happen, when they were near me. It got to the point that every time that two or more people were around me, they would start arguing about minor things.

Then it became more tangible, as I started seeing shadows moving around in my home out of the corner of my eyes, as though a dark figure was walking past. The shadows were so apparent that sometimes I would quickly turn my head and think someone was standing near me. But there never was. Next I started to hear voices. Several times I heard my wife's voice calling my name, but when I would respond, she would say, "I didn't say anything." I also heard my son call my name when he had not. Then I started hearing my name called when no one was home, but me. I tried to pass this off as my imagination, until my wife approached me. My wife is not into the paranormal, but she believes that ghosts or spirits exist and, quite frankly, the subject makes her uneasy. We have a standing joke before I went on an investigation in which she always tells me "not to bring anything home with me." Finally, she asked me what I had brought home with me from Villisca. The reason Kathy said this, was that she was also seeing shadows out of the corner of her eye and was hearing someone call her name as well.

A co-worker and friend of mine, has a daughter who is gifted with strong empathic abilities. Before I went to Villisca for the September investigation, my friend told me of a dream that her daughter had. She said that her daughter told her to tell me to be careful on my trip. In her dream, she saw dark entities around me that could cause me great harm. When I returned from Villisca, my friend's daughter came into our office to have lunch with her mother. I was the only other person in the cafeteria. Her daughter said that she was getting a very negative feeling in the cafeteria. It was making her sick and she stated that she was picking up the feeling from me.

When my friend, and her daughter, came to my 50th birthday party, her daughter sat next to me. She could not eat because she did not feel well. Once again, she said that whatever was causing her to feel ill was emanating from me.

The next week, I was seated at my desk at work when I felt a strong breeze, as though someone had just walked past. I should explain that the building in which I work is a state government office building. The windows are sealed and do not open. When I felt the breeze, I assumed it was a co-worker who had walked into the cubicle that we share. When I turned to see if it was her, I noticed that she was already sitting at her desk with headphones on, working away. When I turned back around toward my desk, I was hit with the awful smell, like hog manure. The smell only lasted for about five seconds. But I knew what I had smelled.

My wife works in the same office as me. Several people who I work with approached her and told her that something did not seem right with me. They said that I looked and acted like a different person. The funny thing is, I was thinking the same thing!

Finally, after several weeks of this, I e-mailed both Paul and Sharon, who had also traveled to Villisca for the investigation. I asked them if they noticed anything unusual since they had returned home. Paul told me that he and his wife were on edge. He also said that they felt depressed. When I contacted Sharon, she said that she decided to separate from her husband after she returned home, but that her decision was unrelated to Villisca. However, everything had been fine before she left.

CHASING SHADOWS

I have a good friend, Lynn, who is a hair stylist that my wife and I both frequent. We went to see her in November of 2008. Lynn has had experiences with haunted houses, so every time that I see Lynn she will ask about new investigations that I have been on. That day, Lynn noticed that something was not right with me. As I was explaining to Lynn that I had just investigated a murder house in Iowa, suddenly she excused herself and went over to talk to my wife. She was speaking to Kathy in such a low voice, that I could not hear the conversation. Later, as Kathy and I were driving home, I ask her what it was that she and Lynn had been talking about in such low voices. Kathy said that Lynn had asked her what was wrong with me. Curious, my wife asked her what she meant by that question. Lynn explained that when she was talking to me, I acted different and that my face suddenly changed and was a pale gray color. She said that I "looked different," and that my face appeared to be that of an old man. It was "not Larry that I was looking at." A few days later, I was driving home alone. I took the off ramp and looked in the rearview mirror to check traffic. What I saw was an unfamiliar face staring back at me. I could see me, but I could also see the facial features of another person, an older man! Needless to say, this was very disturbing! It was probably the most frightening thing that I had seen in my life! I truly believe that there was someone, or something else, in that car with me that night.

193

Seeing someone else's face looking back at me in the mirror was the final straw. I had no doubt that something supernatural was going on. After this incident, I went to a local bookstore in Springfield. I browsed through the section on ghost and hauntings. I found a book about Mary Ann Winkowski, the real life person that the TV series *Ghost Whisperer* is based on. There was a chapter on *Spirit Attachment* in the book. I bought a coffee at the Starbucks inside the bookstore and began reading. I felt like a character in one those TV shows that goes to a library and reads up on how to kill a vampire or werewolf or some other supernatural creature. As I was reading the chapter on spirit attachment, I found a checklist of twenty things that are common in spirit attachments. I was able to check off eighteen of the twenty possible signs.

With little doubt as to what was taking place, I contacted a local psychic medium, from Springfield, Illinois, named Cheryl. I had only met Cheryl one time for a reading, and before that I was a complete skeptic as far as psychics were concerned. But she made a believer out of me. Cheryl is the real deal! I called Cheryl from work and she did not answer her phone, so I left a voicemail message with her. It was short and to the point. *"Cheryl, I have something that I need to speak to you about."* I did not elaborate or say anything regarding what was happening to me.

The next morning, as soon as I arrived at work, my phone was ringing and it was Cheryl. I said hello and she replied, "Larry, this is Cheryl. Where have you been? Have you been to a house or was it a cemetery?" Well, my answer to her questions was both, since we had spent most of our time at the house, but had also gone to the cemetery with Darwin to film the graves of the victims. Cheryl's next question was more like a statement, because she already knew the answer to it. She asked, "Larry, were there murders in the place that you went to?"

Again, my answer was yes!

She had no way of knowing where I had been to. But she described things that sent chills down my spine. Cheryl said, "There were two families murdered in the house, weren't there, Larry?" I responded yes. She then began to describe how she saw two adults murdered and six small children, a total of eight people murdered. She was correct, but she had no way to know this, other than her gift as a medium.

Cheryl proceeded to explain that three, lower-negative entities, had attached themselves to me. When I asked her what she meant by *lower negative entities*, she explained that they are the opposite of angels. I asked her whether she meant they were demons, but she refused to call them that, preferring to call them lower negative. However, I truly believe that she did indeed mean demonic entities.

195

Cheryl explained that I needed to have a spiritual cleansing done as soon as possible, so I scheduled one with her for the following Saturday morning.

I arrived at Cheryl's home at 10:00 a.m. Saturday morning. She took me to the area of her home in which she works with her clients. When she went to retrieve the materials that she uses to perform cleansings, they were not where they would normally be. She explained to me that she always kept her tools of the trade in the same place and that her husband never touches them. They had never been in a different place before. It took her a good twenty minutes to find them and she said that she did not understand how they ended up in a place where she had not placed them. She had the feeling that something did not want her to perform the cleansing.

Once she found the materials, she had me lay down on a table. Cheryl said various prayers while using a Tibetan Singing Bowl, which is a metal bowl that makes a loud and sharp humming sound when hit with a metal rod. She said that negative spirits do not like the high-pitched sound. Cheryl then used a Native American drum to perform some type of a ritual from head to toe over my body.

While she was doing the drumming ritual, I had an overwhelming feeling of sadness overtake my emotions. Tears began to stream down my checks. I could not figure out what was happening to me. After Cheryl was finished, I explained to her about becoming emotionally distraught and she explained to me that I had picked up emotions of spirits in the house and that these emotions were being released.

Normally, a cleansing takes about thirty to forty minutes. My cleansing took over two and one half hours. Cheryl also said that it might be necessary to do a cleansing at my home. Fortunately, that did not turn out to be necessary.

I brought some photographs of the Villisca Axe Murder House for Cheryl to see and, to the best of my knowledge, she had never heard of it before my visit. She pointed out several rooms in the pictures and explained that those rooms had very high spirit activity in them. What was interesting to me was that the rooms she pointed out were rooms in which I had recorded the EVPs.

She continued and told me that in addition to the spirits of the victims and spirits of the killers, something else was lurking in the house that was very evil. Whatever it was, it was inhuman and had never walked the face of the earth as a living thing. She was not sure if it had been in the house prior to the massacre or if the entity had shown up after the murders. She did tell me that I should never return to the house. I explained to her that as a paranormal investigator I felt drawn to the supernatural for a reason and when a place has proved to me that something paranormal is happening, then I must go back to try and find answers.

Cheryl was very understanding and explained that if I must go back; there were precautions that I should take upon arriving at any location, but especially the Villisca House. She explained that the lower negative entities can cause serious problems, including making us physically ill. They can also cause harm to those around us. The precautions included surrounding team members and myself, with the light of God's protection. She added that I should always ask permission before entering any location. When finishing an investigation, I must demand that any entity that is there must remain and is not welcome to come with me. She also suggested backing out of any location, which would help prevent spirit or entity attachment. Cheryl also said that I should use a combination of sea salt and black pepper, called *black salt*, sprinkling it around all entrances and exits of a house or property. She also said to spray rose water on my team members, and myself as negative entities do not like the smell of rose water. I now follow these and other precautions when conducting any investigation.

I returned to the Villisca House for my third investigation on May 11th, 2009 with another investigator, Jamie Sullivan. This was a trip during which I saw the most

CHASING SHADOWS

incredible thing that I have seen in all the time I have been investigating the paranormal.

There were several personal experiences on the May 11th, investigation, including hearing a strange buzzing or humming sound. What was unusual about the sound was that I only heard it in my left ear. Jamie also heard the same sound, but not at the same time. No insects were noted during this investigation that could have caused this sound. The buzzing was more of an electrical or mechanical sound. At about 10:00 p.m., we were downstairs in the parlor of the house, basically just being quiet and listening for any signs of activity. It was a little chilly in the house, about sixty-two degrees or so, and it had been raining. Jamie was tired from the six-hour trip. He decided to lie down and stretch out on the floor. He was wearing a hooded sweatshirt and had pulled the hood up over his head. He had been lying on the floor for only a couple of minutes when he said, "I must look pretty silly laying here on the floor with my hood over my head." No sooner had he said this, when all of a sudden we heard what sounded like the giggling of little girls coming from the downstairs bedroom where Lena and Ina Stillinger were murdered. Jamie immediately rose up, like a vampire rising from its coffin. Excitedly we both asked at the same time, "Did you hear that?" The giggling was loud and clear.

When we checked the room, there were no signs of the little children whose voices we had just heard.

The giggling and buzzing were unusual to say the least, but the most amazing thing that I have witnessed since becoming a paranormal investigator took place during the morning hours of May 12th, 2009. I was upstairs in the children's bedroom. The upstairs is small and laid out as follows: when you head upstairs you immediately arrive at Joe and Sarah Moore's bedroom. If you turn left, walking through their bedroom, there is a wall on your right and a door to your left. If you open this door, you will find a second *half* door. Beyond this door is the attic, where many believe that the killer, or killers, hid on the night of the murders.

Continuing straight down the hallway, you will come to another doorway, which leads into the children's bedroom. That is the room where the Moore's four children were murdered. Straight ahead in the children's room is a bed against the wall. This was the bed where I was sitting at 3:00 a.m. If you sit on the bed and look toward the hallway, you have a view of most everything in which I just described. There are also beds to the left and right of the bed that I was sitting on.

At about 3:00 a.m., it was pitch black upstairs. The house does not have any electricity and the only light was coming from a battery-operated lantern that we had left in the parlor downstairs. Jamie decided to take a brief nap, so he was sleeping on the bed to my left. We had been monitoring the upstairs since about 1:00 a.m. According to legend, at 2:00 a.m. a train will go through town and when it blows its whistle, a residual haunting is triggered. A fog or mist will go from room to room, coming up the stairs following the path that the murderer(s) allegedly took.

To verify the legend, we set up a K-II meter and another EMF detector on the landing of the staircase like an alarm system, figuring that the detectors would alert us if anything passed the EMF detectors and was giving off an electrical of magnetic charge. We also had a video camera set up in the Moore's bedroom facing the landing of the staircase, hoping to catch this residual on video. We also had another video camera recording in the attic.

At 2:00 a.m., the train whistle blew right on cue, but nothing happened. About forty minutes later, the battery for the camera in the Moore's bedroom went dead, although this was nothing paranormal. We had simply failed to bring the charger for this particular camera. Consequently, the only camera filming at the time was the camera in the attic. At about 2:45 a.m., Jamie decided to lie down on the bed to my left to take a brief nap. I continued to monitor the EMF detectors and listen for any noises. Darwin Linn, the owner of the house, had told me that there were twenty-eight trains that passed through Villisca each day. At 3:00 a.m., another train whistle blew. All of a sudden, I observed a neon green light coming up the staircase. It appeared as if someone was carrying a lantern. The light was very bright and lit up the wall. What was especially strange about this light was that it acted almost like a fog that blanketed and illuminated everything it passed.

The light continued to the top of the stairs, and then, just like a person, it turned left, first lighting up the ceiling of the Moore's bedroom, then the wall, until the entire bedroom was glowing brightly in this green light. There were no orbs and no source of illumination that I could see; it was simply pure light. The green light then made a right turn, came out of the Moore's bedroom, and headed down the hallway toward me. The color of the light was a color I have never seen before. I can best describe it as a combination of neon green and the shade of green that you see when looking through night-vision goggles.

The light was about parallel with the door that led to the attic when I yelled, "Oh, shit!" and then called to Jamie to "wake up and see this!" By the time that he woke up and got to his feet, the light vanished. To this day, I wonder what would have happened if I had not yelled for Jamie. Would the light have kept moving toward me? All I could think about for weeks after was the light, trying to figure out what it may have been. The entire event lasted ten to fifteen seconds, but it left a memory that is forever imprinted in my mind. Because of this light, the giggling, and the EVPs, I made a return trip in October of 2010.

October 10th, 2010, was my fourth trip to the Villisca House. I brought along two other investigators, Chris Mason and Janet Morris. We had several personal experiences that night, including the temperature rising from seventy-two degrees to eighty-nine degrees after we asked whatever was in the house to make the temperature rise. Right before I asked the question, Chris, who had been sitting on a bed in the Moore children's bedroom, moved over to another bed that Janet and I were sitting on. Chris said he had a feeling something evil was standing by him, so he moved over by us. At that point, I asked whatever was in the house to make the temperature rise. When I did, Janet became sick to her stomach and had a feeling that something very bad was in the room. She said that something was telling her that she should leave, so I had Chris take her outside.

Just as Chris and Janet were getting ready to leave, the temperature spiked from seventy-two degrees to eighty-nine degrees in only two or three seconds. As they went outside, I stayed in the house and changed the DVD in the video camera I had left upstairs. After changing the videodisk, I went back downstairs. Knowing that whatever resided in the house had probably followed me home in 2008, I said, *"Okay, I'm alone in the house, now is your time to prove to me that you are real."* As soon as I said this, I heard three loud footsteps coming from upstairs, which seemed to walk from one side of the floor to the other. It was a loud *boom, boom, boom.* Shortly after this, Chris came back in to check on me. He said I was acting strange, so we should go outside and take a break. Janet decided to stay outside for the remainder of the night and never returned to the house.

The next day, I arrived home in Illinois about 3:00 p.m. When I arrived, both my wife and son were gone, so I decided to check my email in my home office. I was sitting at my desk reading emails, when all of a sudden in the corner of the room, I heard a *hissing* sound like a large snake would make. I got up from my desk and checked the corner. Nothing was there. I sat back down and continued reading my email. Suddenly, I heard loud clawing and scratching off to my right in a cabinet. I immediately got up and checked the cabinet, but nothing was there either.

After a few minutes, I heard something moving around in our bedroom, which is right next to my office. When I went to check it out, again, there was nothing. Irritated, I sat back down at my desk. Moments later, three loud footsteps echoed down the hall next to my office. It was exactly like what I had heard in Villisca. Then, I heard a voice, which was not like my own. It said, *"You don't have to come to me, I can come to you!"* This really shook me up. Later that night, I heard someone in our kitchen. I would hear the sound of glasses or pots and pans. But when I would check, no one would be there. This went on for a few days then stopped. My wife also heard things, but I did not let her know that I was hearing them as well, because I did not want to alarm her.

When I reviewed the audio from the Villisca house after the October investigation, I discovered that I recorded some interesting activity. At one point in the evening, when we had all decided to go outside and take a break, the three of us can be heard leaving the house. A few moments later, you can hear what sounds like a barroom brawl, complete with chairs being dragged and moved. Of course, nothing had been moved or disturbed.

I still have over seventeen hours of audio clips to review from the Villisca house, and I believe there are probably EVPs on them, but I stopped reviewing evidence from Villisca because every time I start to listen, it feels like someone is standing behind me. It is such an overwhelming feeling, that I will quickly turn around expecting my wife or son to be standing behind me, but there is never anyone there.

Investigation Summary

Without a doubt, the Villisca Axe Murder House is one of the most haunted locations that I have visited. I have heard the laughter of children in the home, so I believe that at least some of the Moore family may still reside there. But I have no doubt that something else more sinister lurks in the shadows of the house that seems to cause an overwhelming feeling that you are not safe or welcome there. I believe that what is in the house may be as Cheryl the psychic said, something that has never walked the face of the earth as a human! What happened to me has happened to other as well, but it doesn't seem to happen to everyone that visits the house. So why did it choose me?

I am not sure when I will return to the Villisca house, but I know that as long as the house is there, someday I will go back. Is the veil to the other side thinner at 508 East 2nd street, which allows us to experience things that other places do not seem to allow? Or is it whatever that lurks in the house, allows us to experience the terror that it has to offer. I crossed paths with the Villisca house for a reason. Maybe it was to show me that the supernatural exists, that it is not simply fantasy or make believe. It definitely taught me that spirit or entity attachment is real and can be dangerous. What would have happened if I didn't know Cheryl? Would things have become so bad that there would have been no hope? Maybe something or someone on the other side, is watching, but whom or what and why?

The Villisca house has taught me that evil exists and may lurk around the next corner, so we must keep up our guard at all times. It has also taught me that the supernatural can cross over and exist in the physical world. I have seen many things that I cannot explain in the eight years that I have been investigating the paranormal, but nothing compares to the experiences I have had at the Villisca Axe Murder House. If you decide to investigate this location, beware, because this house is the real deal and it may choose you for its next intended victim. So beware, and don't let your guard down, because your life and happiness may depend on it!

Cumberland Sugar Creek Cemetery

CHAPTER NINE

Cumberland Sugar Creek Cemetery
Glenarm, Illinois

At the beginning of this book, we began our adventure at a strange rural cemetery in Illinois, explored several haunted historical buildings in Missouri, and then traveled to Iowa to investigate the location of an infamous murder. Finally, we return to Illinois to complete our journey in another rural cemetery. I saved this location for last, because the events you are about to read happened in June

2011, just a few weeks before the original release of this book.

This chapter is about an encounter that I experienced and still cannot explain. Fortunately, or unfortunately as the case may be, friend and fellow paranormal investigator, Chris Mason, also experienced the encounter. If nothing else, we can reassure each other that what we saw was real. As you will see, the events leading up to our strange encounter in Cumberland Sugar Creek Cemetery are almost as bizarre as the event itself.

It began when I attended a local paranormal meet up at Lincolnland Community College in Springfield, Illinois. The meet up is called the *Prairieland Paranormal Consortium*. My friend, and a fellow paranormal investigator, Carl Jones, founded it. Carl serves as the host for this monthly meeting. He is also the founder and host of another group called the *Central Illinois UFO Group*, which also holds monthly meetings at the college. The Paranormal Consortium is an open forum for those interested in all things paranormal, while the Central Illinois UFO Group normally limits its discussions to UFO related activity. However, many of those who study and investigate the supernatural have come to the conclusion that all things paranormal may be related in some inter-dimensional way, so discussions occasionally stray into territories other than strictly UFOs.

During our meeting on Saturday, June 25th, 2011, Stan Courtney presented photographic evidence of possible Bigfoot activity between the communities of Glenarm and Chatham, Illinois.

Stan showed several photos of an eighteen inch long by seven inch wide footprint that was taken underneath an apple tree, on private property, near Covered Bridge Road in Chatham. The family told Stan that they never once picked any apples from the tree, but the apples were always gone. The family also reported hearing strange noises in their backyard. Among Stan's photos was that of a rabbit carcass laid out underneath the tree. The gruesome thing about it was that the rabbit's head had been pulled off. Stan explained that in many case reports of suspected Bigfoot activity, carcases of animals often appear on the property where things such as fruit, vegetables or other animals have been taken. Researchers believe that this is a type of *gifting* from Sasquatch to show thanks or to repay for what they have taken.

I was involved in an incident in the fall of 2010 that gave credibility to the Chatham footprint. There are approximately 700 employees in the building in which I work, and many of them know that I am a paranormal investigator. One particular friend and co-worker asked me to come out to the parking lot to look at a handprint on the rear window of her van. The rear window was very dirty, since she lives in a rural area, where there are both dirt and gravel roads. The area in which she lives is approximately eight miles from where the Chatham footprint was found. When I went out and looked at the handprint, it appeared to be approximately ten inches long, resembling that of a primate. The features that most stood out were the human-like lines on the palm, but the handprint appeared to have long, hooked claws or fingernails. It was like nothing that I had ever seen before, that is, until Stan showed a picture of the Chatham footprint. The toes on this foot had the same long hooked claws or fingernails, just like the handprint on the back of the van.

During Stan's presentation, a member of the paranormal consortium spoke up and said that just the night before, on Friday, June 24th, a local paranormal group had been investigating the Cumberland Sugar Creek Cemetery between Chatham and Glenarm, Illinois. The group was seated in the cemetery conducting an EVP session, when all of a sudden something or someone began throwing rocks and dirt clods at the group. One of the group members was struck in the cheek and her face was cut. Rock throwing is considered classic Sasquatch activity, and is meant to scare off intruders from their territory.

Strange footprints were also discovered under the covered bridge on Covered Bridge Road. Stan investigated the prints that were discovered there and believed them to be authentic. As far as he knew, the people who found the tracks did not know each other. Just down the road from where the tracks were found, another man reported that his dog was startled by something in the forested area near covered Bridge Road and was staying close to home rather than running in the woods as he normally would.

During the meeting, Carl Jones mentioned how he had recently seen a round, spherical object hovering by a pond near the Glenarm exit, off Interstate 55. The sphere was round and shiny. He could tell it was not an airplane or helicopter, because it did not have wings like an airplane nor did not have a rotor blade like a helicopter. It was there a few seconds then was gone.

Carl works at Lincolnland College and was there the following Monday, June 27th. During the middle of the morning, Carl and several co-workers went outside to take a break. While outside, Carl looked up and saw a cylindrical object flying in the sky. The odd thing was that it did not have any wings. He pointed it out to three other people who were there.

This is where the story gets a little weirder. First, let me say that I have known Carl since 2006 and have always known him to be honest and straightforward, so when Carl tells me something, I believe him.

On Tuesday, June 28th, Carl was at work when he received a phone call from a fellow member of the Mutual UFO Network (*MUFON*), who told him that a wingless, cylindrical object was seen in the sky in northern Illinois. The object was heading south toward Springfield and that if Carl went outside, he might be able to see it. Carl went outside to take a look, but did not see anything unusual, so he went back inside. He returned outside a short time later to take another look. This time, sure enough, Carl saw what he described as a wingless, cigar shaped object, but now it was travelling in an easterly direction at a fast, but consistent rate of speed. He watched the object until he lost sight of it.

I subscribe to an internet paranormal blog that covers the gamut of the strange and unusual. A few days later, the blog reported that the same morning Carl saw the cigar shaped object, identical objects were also reported in the sky near Kankakee, Illinois and Kokomo, Indiana. The descriptions were identical in all four sightings. On Tuesday evening, June 28th, at the monthly Central Illinois UFO Group meeting at Lincolnland Community College, Carl discussed the sightings that he had recently witnessed, including the metallic sphere that he had seen near the pond at the Glenarm exit off Interstate 55.

During the meeting, Chris Mason and I discussed the possibility of heading out to Cumberland Sugar Creek cemetery, to see if we could determine where the paranormal group had been sitting, when they were pelted with stones, that were thrown from somewhere near the fence line. On the way to the parking lot after the meeting, Chris, Carl, local MUFON director John Jenner, and I continued our conversation regarding the strange events that had been occurring in central Illinois. Carl explained how his email account had recently been hacked. He has a Facebook page for the Central Illinois UFO Group, which is frequently updated with various topics about UFOs and UFO sightings, and we were discussing the possibility that someone checking up on his activities may have compromised his email account. At one point during our discussion, the security light that we were standing under went completely black, but all of the other security lights in the parking lot remained lit. As we were finishing up our discussion and were ready to leave, the security light once again lit up. Was this a coincidence? Maybe, but it was an eerie one to say the least.

Neither Chris nor I had ever been out to Cumberland Sugar Creek cemetery, so after our parking lot discussion about Carl's hacked email account, Carl gave Chris and I general directions as to how to get out to the cemetery. We left the parking lot and I drove my car to a McDonalds located near Interstate 55, just a couple miles from the campus. Chris followed me and I left my car in the parking lot. We then headed to Glenarm in Chris' vehicle to see if we could find the cemetery.

I was not familiar with reported activity at Cumberland Sugar Creek, but later found out that local authorities in years past would occasionally stake out the cemetery looking for underage drinkers. I was told that several times that the authorities had reported seeing lights in the cemetery. Thinking that kids had caused the lights, the police would investigate, but would always come up empty handed.

Carl's directions were very accurate. Chris and I found the cemetery without any problem. However, because we came straight from the college, we did not have our paranormal investigating equipment with us. More importantly, we did not have a flashlight either. The moon that night was a waning crescent, which is only about one sixth of a moon. The sky was clear for the most part, but the cemetery was very dark. We arrived a few minutes before 10:00 p.m. Once we got out of the car and walked a short distance. Our eyes seemed to adjust to the darkness and we could make out objects or obstacles, such as tombstones, that were in our paths without difficulty.

Just a few minutes later, I glanced up toward the sky and noticed an object about the size of a very bright star heading west to east. I pointed out the light to Chris and asked if he thought it might be a satellite. Neither of us was sure as to what it was, but it did not have flashing lights like a normal aircraft would. I turned away for a moment to check out the cemetery, because I thought I saw something moving. No sooner had I looked away when, Chris yelled, "Hey, the light vanished!" I looked up and, sure enough, it was gone.

A few moments later, we saw another, similar light coming from the southeast, and heading in a northeasterly direction. A few seconds after that, we saw another light coming out of the south and heading north. This time, I kept my eyes focused on the light. What I noticed was that both of the lights seemed to be headed in an upward trajectory. As I watched, the lights became fainter and fainter. Then it hit me: the lights were not extinguishing; they were heading so high into the atmosphere that they were disappearing from sight. I do not know of any military or commercial aircraft that travels that high, so I was completely baffled as to what we were watching.

After the lights disappeared, we scanned the sky for a few more minutes, but did not see anything else unusual. Our focus returned to the cemetery and our mission of trying to figure out where the paranormal group from Springfield had been sitting, as well as the direction from which the rocks and dirt clods, that they were pelted with, may have come from.

By this time, our eyes had adjusted to the darkness and we could see our way around the cemetery without stumbling over any obstacles in our paths. There were several times when both Chris and I thought we saw movement in the cemetery, but wrote it off as our eyes playing tricks on us. The temperature seemed like it was in the low eighties, but since we did not bring any equipment with us, I was not able to determine the exact temperature. Both Chris and I walked into cold spots that caused icy chills and goose bumps. Drawing our conclusions from the layout and terrain of the cemetery, we explored along the fence line in the southeast corner of the cemetery, because we felt that was probably the area where the Springfield paranormal group had been, when they were pelted with rocks. It was hard to tell in the dark, but the fence line was not as heavily camouflaged as we thought it would have had to be for anything as large as a Sasquatch to be hiding behind.

On a second excursion to the cemetery the following week, we were able to determine in the daylight that just beyond the fence line to the south, the terrain sloped down so that it could have been possible for a large person or creature to have squatted down and thrown rocks without being seen, especially under the cover of darkness. We spent a good ten to fifteen minutes exploring the fence line from the southeast corner of the cemetery to the northeast corner and back.

We had just walked back toward the south end of the cemetery. We were approximately forty feet from the fence line on the east and another thirty to forty feet from the fence line to the south. Chris was standing approximately seven to eight feet in front of me, facing south. I was also facing south and was behind Chris, at an angle that was a step or two to his left. We were again discussing where someone could have possibly been standing while remaining hidden from the Springfield group, when all of a sudden, Chris and I were lit up like Christmas trees and I could see Chris as plainly as if it were the middle of the day. My first reaction was to turn my head and look behind me. As soon as I did this, my attention was drawn upward. When I turned and looked up, I saw an extremely bright *yellowish-white* light. It was so blinding, that I extended my arm and hand toward the light, to shield it from my eyes. Chris reacted in the same manner as I did. I am still uncertain how high above us the light was, but it was at least eight feet or so. The light surrounded both Chris and I for an estimated twenty foot radius. The blast of light lasted for about four to five seconds.

The light disappeared as quickly as it appeared. It was there one moment and gone the next. When I looked up, all I could see was the brightness of the light, like in those old spy thrillers where a blinding spotlight is shinned into someone's face to get him or her to talk. There are no security lights inside the cemetery, or any streetlights around it. As a matter of fact, the only lights in the cemetery were a couple of small solar lights that had been placed on the ground near some of the graves at the far end of the cemetery. Chris and I both turned toward each other. Chris tried to speak, but the only word that made it out of his mouth was, "What?" He then immediately began to choke. He leaned on a tombstone that was between the two of us and he was unable to speak for several minutes. Once he was able to talk, he explained that his throat felt as if someone was holding a lighter against it. He compared the burning sensation to a bad case of acid reflux.

Chris and I swapped descriptions of what we had seen and how we reacted. Our stories were identical. I asked Chris to check his watch to see what time it was. During my research, I have identified events where people have had similar incidents. In some of these, there have been cases of missing time. Chris' watch has a light on it, so he was able to see the time. It was 10:25 p.m. So, given the three or four minutes that it took for Chris to be able to speak, the light hit us sometime around 10:22 p.m.

There are a couple of puzzling things about the incident. First, because we were illuminated by a strange light out in the middle of a dark cemetery without flashlights, one would think this would have been ghost or spirit related. But from the very beginning, neither Chris nor I had the feeling that it was ghostly at all. It had the feeling of being something otherworldly, or possibly inter-dimensional, but not spiritual. If the light had come from the ground, or even from the sides, maybe we would have felt it was ghostly in nature, but this came from above. When I saw the neon green light on the second floor of the Villisca Axe Murder House, it totally had a ghostly or spiritual feel to it. The light in Cumberland Sugar Creek Cemetery was nothing like it.

Our reaction to the light was also puzzling. We were initially excited about what had just taken place, but that feeling only lasted a few minutes. We continued to look around the cemetery as if nothing happened. It was almost like we had been numbed to the experience. It was not until we were in Chris' car heading away from the cemetery that it hit us like a sledgehammer. As we drove away from the cemetery, we both gasped, "What the hell just happened?"

The excitement for the night was not quite over yet. Chris had his cell phone with him, and after being in the cemetery in complete darkness for about forty minutes, he remembered that it had a flashlight on it. We turned back and continued our exploration of the cemetery. We returned to the far southeast corner of the cemetery, where there was a mound of soft dirt piled in the corner next to the fence line. Chris is a hunter, and so he was shining his light on the ground looking for deer tracks. Instead, he found three footprints. Footprints in a cemetery are not necessarily unusual, but these were barefoot; we could see the toe and heel prints. What drew our attention even more was the size of the footprints. I wear a size eleven and a half shoe. When I placed my foot next to the print, the print was three to four inches longer than my foot, which would mean that the footprint was at least fourteen to fifteen inches long. We found two more footprints nearby, which were also barefoot. Chris took several photos with his camera phone to document the tracks. The tracks appeared to have been there for several days, which meant whatever made the tracks could have been in the area the night that the Springfield group was pelted with the rocks and dirt clods. We were only a mile or two from where Stan Courtney had conducted his investigation and took the photo of the larger, eighteen inch track.

After finding the footprints, Chris and I made one more walkthrough of the cemetery and then decided to call it a night. We agreed to come back another time with flashlights and our equipment.

As Chris was driving back to my vehicle, what had happened to us finally sunk in and we started to become excited about it. As we discussed the event, we both had the feeling that something had taken a picture of us or had scanned us for some reason. It was as if something had shinned a big spotlight on us, but for what purpose? Did something want to let us know that it could see us, or was something trying to scare us? I did not get the feeling that something was trying to scare us off, because we had remained calm shortly after the event and did not feel any real concern until we left the cemetery.

I have learned from my experience as a paranormal investigator that when you are in places such as a cemetery, haunted house, or some other type of a dark or secluded location where visibility is not the best, your other senses seem to become sharper and more focused. When Chris and I were in the cemetery, it was very dark, so our hearing seemed to pick up even the slightest of sounds or noises. Just before, during, and after the encounter with the bright light, we did not hear anything unusual. The light caught us completely by surprise.

After leaving the cemetery, Chris drove me back to my SUV, which I had left in the McDonald's parking lot near Interstate 55. We discussed what had happened for a few more minutes, and then we each headed our separate ways.

A few minutes later, I called Chris on my cell phone to continue our conversation about the events of that evening. We were probably about fifteen miles from each other and both of us were driving on Interstate 55, but heading in opposite directions. Interstate 55 is in range of many communication towers and both of us were in good reception areas at the time of my phone call. We had been carrying on our conversation, without interruption, when all of a sudden; I could hear what sounded like a distorted mechanical voice on the other end of the call. This went on for forty-five seconds to one minute. I kept asking Chris if he could hear me, but the strange mechanical voice continued to persist until I decided to hang up and call back. I have an old cell phone and, normally, if I am in a low area or an area where cell phone towers are few and far between, my phone will simply drop the call. But with this call I continued to get that distorted, mechanical voice. I called Chris back, and he explained that he had heard the exact same thing on his end of the phone.

Cumberland Sugar Creek Cemetery

When I returned home, I explained to my wife Kathy what had happened with the light and then, because it was so late, I went to bed. During the middle of the night, I woke up and was sweating like crazy. I felt like I had a fever, even though our air conditioner was turned on. The next day, I was sick to my stomach all day. For the two days following the event at Cumberland Sugar Creek Cemetery, I was extremely tired and lethargic. I work out on a regular basis and do not fatigue easily. I spoke to Chris a couple of days after the incident and he said that he also felt more tired than usual.

Since then, Chris and I have discussed the incident many times. We both have come to the conclusion that the light source did not come from inside the cemetery. We had the feeling that whatever it was came from somewhere else, but from where? Was it extraterrestrial in nature? We had not seen or heard anything, but a light just does not come out of nowhere and for no reason. It felt as though we were being looked at under a microscope, but why? And how did anyone or anything know that Chris and I were there? It was so dark that no one would have been able to see us. One important thing to note is that the object Carl had witnessed hovering over the pond, near the Glenarm exit, would only have been two miles, *as the crow flies*, to Cumberland Sugar Creek Cemetery. Carl had described it as being roundish in shape and shiny. I wonder if the reason Carl could see the object in the daylight was because it was reflecting the sun, and whether it would have been visible at all if it were dark.

The following Wednesday, July 6th, 2011, Chris and I returned to Cumberland Sugar Creek Cemetery with our equipment and flashlights in hopes of having a second encounter with the strange light. Unfortunately, there was very little excitement or activity. I did have one strange experience at about 8:00 p.m. It was still light out and I had been over near the fence line on the southeast corner of the cemetery. Chris was about thirty yards away from me. When I got to within about ten feet of Chris, I heard and felt what sounded like a large bird fluttering wildly right behind my head at about neck level. I could even feel the breeze from the flapping, and I definitely heard the sound of wings. I turned around quickly, arching my back, and pulling my shoulder away in surprise. It was as if I was trying to get away from something and expecting to see this large, angry bird behind me.

When I turned around, however, both the noise and the breeze stopped. Nothing was there. I quickly turned back and looked at Chris, because I knew that he would have had to of seen it, whatever it was. Chris had a puzzled look on his face, like he did not know what was wrong, but he could tell by the look on my face that I was going to ask him if he had seen anything. He immediately said, "Don't ask me, there was nothing there!" Then added, "I was wondering what the heck was wrong!" He said that it appeared as though I was trying to get away from something. I explained to him what I had heard and felt. He reassured me that there was definitely nothing there.

This turned out to be the only activity that night, and since we both had to get up early in the morning to go to work, we headed home.

Investigation Summary

Chris and I had planned to go back for a follow up investigation in the near future, but unfortunately, news of the possible Bigfoot evidence had created local media frenzy. Because of the publicity, more and more people have taken an interest in the Cumberland Sugar Creek Cemetery and the Covered Bridge Road area. The local authorities have already tossed a couple of investigators from the cemetery. In my opinion, when a story such as a Bigfoot sighting hits the media, any ongoing investigations are compromised from that point on, and the environment where the sighting or evidence was reported is no longer a controlled environment.

So, in summary, is Cumberland Sugar Creek Cemetery haunted? My best answer to this is, I do not know! I say that

CHASING SHADOWS

I do not know simply for the fact that I believe something truly paranormal happened to Chris and me, but it did not have the feeling of being spirit-related. A skeptic would say, "How can you call what happened paranormal, when you do not even know where the light came from?" My answer to this would be, "That is exactly my point." The light appeared out of thin air. Lights appearing out of thin air are not normal. Still, even though we were in a cemetery and one would normally think that a strange light suddenly appearing in a cemetery would be spiritual or ghostly in nature, we did not, because this light came from above us!

When the phenomena occurred, it illuminated a good twenty-foot radius around Chris and I. It caused me to think that whatever caused the light had a lot of energy. However, I did not get that eerie or heavy feeling that I normally associate with having a spirit present. Most always, when I have experienced spirit activity, such as the time I was punched in the back at Anderson Cemetery or when I have seen and heard things at places like the Villisca house, I have always had the feeling that someone or something was there. It is that feeling that comes over you that something is not quite right, a heaviness that causes an anxious feeling. None of that happened when the light hit us. The only effects were those that I experienced later that night and over the next couple of days; such as feeling feverish, sick to my stomach, and fatigued.

As for hearing what sounded like wings flapping directly behind, then turning around and not seeing anything, I have no explanation for that either. But I will say that this type of *activity* is more like what I would have expected to have happen in a cemetery investigation.

Based on the experience that I had with the light and on my gut feeling alone, my opinion as a paranormal investigator is that one of two things happened. One theory is that we were just in the right place at the wrong time and experienced some type of an inter-dimensional anomaly. Alternatively, it could have been something otherworldly and possibly extraterrestrial in nature, and Chris and I were either tagged or scanned by something with intelligence. If the latter is the case, I wonder who tagged us, where did they come from, and what did they want? If we were somehow tagged, does that mean that we are being monitored from a distance, or does it mean that someday they will be back?

I plan to return to Cumberland Sugar Creek Cemetery for further investigation of the activity, but probably will wait until the Bigfoot stir dies down a little. Right now there are too many experienced, and inexperienced, investigators roaming the Covered Bridge area trying to conduct investigations. One thing is for sure, you can bet that in the meantime I plan to keep my eyes to the sky and will be looking over my shoulder, because one day I may come face to face with whatever was monitoring us that night and find out firsthand what they want!

CONCLUSION

Advice for Paranormal Investigators

By reading this book, I assume that you have a strong interest in ghosts, hauntings and things that go bump in the night. Whether your interest lies simply in reading about the supernatural or whether you plan to investigate the paranormal for yourself someday, the most important thing that I can convey to you is my sincerity when I tell you that the supernatural is very real and can also be dangerous. The stories in this book are not legends, they are all true accounts of things that I have experienced and encountered firsthand in my years of investigating.

The hours that I have spent waiting, staking out cemeteries and old buildings, during all-night investigations, hoping to catch that one piece of evidence that will prove that the supernatural exists, may seem like a waste of time to some. However, when you see something or have that personal experience that makes you stop and shake your head at what you just saw or experienced, or that causes you to question some of the things that you were taught to believe. Like the green light I witnessed at the Villisca Axe Murder House, or the time that I was punched in the back at Anderson Cemetery. It makes it all worthwhile and keeps you coming back for more. It opens your eyes to a whole new world that has been around us the entire time.

Believe me, you will never forget the feeling that you had, the first time you heard your name being whispered, by a voice that you recorded, when you knew that no one else was there.

My advice to those of you reading this, and who are new to paranormal investigating, is to never let your guard down and to always *expect* the unexpected. What we are delving into is a complete unknown. It is a field where there are no experts, but one in which there are those who have experienced the supernatural firsthand and who have had personal contact with things they may not understand, but have experienced nonetheless. We must learn from our experiences, and the experiences of others like us, and take note of the circumstances and conditions in which supernatural events occur.

One of the most important things to remember, when setting up an investigation, is to always keep as controlled an environment as possible, by minimizing the number of investigators and third party on-lookers at the site. More is not better when it comes to this. It is much easier to review audio evidence when a small number of investigators are on site. By limiting the number of people allowed at a site, it is much easier to keep track of where people are and to keep conversations limited. We must scientifically test our theories and the theories of our colleagues. In order to do this, we must eliminate unnecessary factors such as noise artifacts, caused by too many investigators. I have been on investigations with other groups, who take six or eight people into a small house. When it came time to review my audio from these investigations, there was so much chatter and talking by investigators that I could not, in good conscience, call any odd sounds or voices captured on the recorder *evidence*.

Furthermore, always heavily scrutinize the evidence that you review from an investigation. Make sure that you have eliminated all possible logical explanations before calling any piece of evidence paranormal. As the saying goes, when in doubt, throw it *out*!

Through the process of elimination, trial and error, as well as by comparing evidence, shared from trusted sources, with evidence that we have obtained, we are able to scientifically evaluate the data that we have recorded. Ghosts, spirits or supernatural beings must operate by some principles and laws of physics, but not necessarily the ones that we have been taught or are aware of. So, as investigators, we must think outside of the box when trying to figure out our supernatural experiences.

As paranormal investigators, we are the pioneers of a new frontier. Testing the waters and navigating through unchartered territories, attempting to lay the foundation for others to build on. If we build a strong enough foundation, we will someday be able to break through the barriers that prevent us from establishing that the supernatural exists. Proving once and for all, that life continues when our physical bodies give out.

Never forget that even though the majority of haunted places that you investigate will be uneventful or even mundane at times, you can never know what you might encounter, and you will usually encounter it when you least expect it.

Always prepare yourself by surrounding yourself with the light of God, and say a prayer asking for God's protection. Be respectful of the places and the spirits that you encounter, but before you leave any location, verbally demand that any spirits that are present must either remain at the location or go into the light, but that they are not welcome to come with you. But remember, just by demanding that the spirits either stay at the location or go to the light, does not mean that they will follow your wishes. When you go home after an investigation, always cleanse yourself by taking a sea salt bath, asking that God lead any *lost souls* to the light. After bathing, always back out of the tub so that the spirits do not reattach themselves. Let the water go down the drain and the spirit energy go with it. I always wear a St. Michael the

Archangel medallion. I don't pray to the medallion, but it gives me strength and reminds me that higher beings do exist that watch out for us and we can call upon them if they are needed.

Now is an exciting time to be a paranormal investigator. It is much like during the time of the discovery and perfection of the microscope. By looking through a polished piece of glass, a whole new living world was discovered. A world that always existed, but was so tiny that without the invention of the microscope it would have remained hidden and undiscovered forever.

As a paranormal researcher, I am searching for that piece of equipment that will turn out to be the microscope to the supernatural, one that will uncover and unveil a world that exists around us all the time. If you think that you have privacy, think again. Are we really alone when we take that shower, or bath in the privacy of our own home, or are we the specimen under the microscope being studied by our brethren from the other side or another dimension?

As paranormal research methods and equipment become more specialized, we are finding that additional evidence is being obtained. We must figure out a way to hone in and filter out the white noise and artifact, which distort and cloak the voices that we record. We must also find a way to filter the light and darkness, which prevents us from videotaping the things that exist on the other side, so that we will start finding the answers to our questions. If we can figure out which frequency is the actual line of communication between other dimensions and that of our own, we will be able to clearly communicate with these other life forms.

Many believe, and I agree, that for some reason over the last few years the veil between the living and the dead has become thinner.

Spiritualists, metaphysicists, as well as those that follow the teachings and the prophecies of the ancient Mayans, Hopi Indians, and other ancient civilizations, believe that we are in the beginning phase of a great awakening. That this awakening is due to cosmic alignments, which are in the initial phase and will only get stronger in the next few years. It is believed that this cosmic alignment is increasing the power and strength of the magnetic fields on earth. As we have come to learn in our paranormal investigations, more often than not, when there is spirit activity, or when supernatural things occur, we normally detect higher magnetic and electronic interference. My theory is that this increase in magnetic energy somehow either thins the veil between the physical world, spirit world, and other dimensions; or possibly increases our ability to detect and even see spirits or supernatural beings.

The Supernatural

I know that I have thrown the word *supernatural* around in this book quite often, so I should probably explain to you what my belief is; regarding this phenomena. I believe that it is a state of existence, not an equal existence, but a higher existence. I also believe that it cannot be explained or understood by using logic, based on our natural laws of physics, because the laws of physics of the supernatural are more highly evolved than our natural and physical laws.

We, as humans, cannot defy the natural laws of science and physics without consequences, but I absolutely believe that the supernatural can interact, and even override, the natural laws of our physics. I also believe it is one of the reasons that supernatural things can exist all around us, without our knowledge. So if this theory is correct, then why can we sometimes see, hear, or interact with supernatural things and events? My answer to this is that all things and levels of existence come from the same source. Since we come from the same source, or Creator, it would serve to reason that some of the laws, which control our physical and spiritual existence, would logically be the same or similar. I also think that, in the grand scheme of things, one of the primary purposes of our existence is to learn, and one way

CHASING SHADOWS

that we are nudged into to learning is through our natural curiosity. So for this reason I believe certain environmental conditions purposely exist that allow us to get a glimpse of other realms of existence, like a sneak preview to prepare us for future events to help us continue on a path that allows us to evolve and learn.

To me, supernatural and paranormal events are one and the same. They are things that happen due to an interaction between different realms of existence, both by design and by accident. I believe that there is one existence, an existence that is so massive that it cannot be measured. It cannot be measured because it is impossible to measure something that we cannot comprehend. This one existence has multiple realms of life that are separate, but sometimes become intertwined.

I believe that this one existence is constantly expanding and that it expands because of the thoughts of the collective consciousness, which you and I are part of. The world's population is estimated to reach seven billion by October 31st, 2011. So just imagine the vast amount of expansion that would be created each day if you multiplied your thoughts by seven billion thoughts a day. That number only accounts for the collective thoughts of one planet in a universe that is incomprehensively large. We have no idea how many other civilizations may exist in other worlds both physical and non-

237

physical. It was estimated in 2002 that over 100 billion people have died on this planet since the beginning of time. If we have eternal life and eternal thoughts, multiply the expansion of our existence by another 100 billion thoughts per day.

The expansion of this existence is so vast that we will never get to see its entirety or get to the point of seeing, experiencing, and completing everything that there is to do. Because if we did, that would be the end of the line for us and eternal life would stop. Our existence must continue to expand in order for *time* to continue. I also believe that we are living at a planned timetable, which keeps us on a steady pace, which will not allow us to catch up to the expanding cosmic consciousness, like a greyhound chasing the rabbit at the racetrack. This divine plan allows life to continue forever. I believe that there are the divine beings that oversee all that exists that guide us and direct us when we temporarily lose our way.

Where to Go Next?

On August 24th, 2011 my team from Urban Paranormal Investigations and I will be investigating a place known as Black Moon Manor in Greenfield, Indiana. This will be my first paranormal adventure to the Hoosier State and I am looking forward to finding out what this place may have in store for me.

John C. Estes constructed the manor, which is located in the heart of Indiana and in the middle of Buck Creek forest, in 1859. It was the first house built in Hancock County, Indiana. Located at the top of a hill, the manor gives you a feeling that it knows who is coming to visit.

The house has a history of tragedy, and many of its former occupants have met their fate in this unforgiving estate. As many as 200 or more died in the house after it was converted into a small pox hospital during the outbreak of 1892. Men, women, and children perished during the outbreak and were buried in a cemetery behind the house.

During the outbreak, there was only one doctor in the local community to care for the sick. He was owner of the Manor. He allowed those who had been infected with small pox to stay there so as to keep a vigilant watch over the sick in hopes that the disease could be contained. The numbers of the infected were too great for one man to handle alone, however, and many of the sick died on the property.

To help prevent the spread of the outbreak any further, the dead were buried behind the home in a private cemetery. Due to the rate of people dying at the manor, an addition was built and a mortuary established to prepare the dead for burial. The mortuary was complete with a preparation room, funeral parlor, and cold storage area in the basement where the bodies were held until their burial. Many of the tombstones still stand in this small cemetery today.

The current owner of the house, Matt Speck, found it by accident while looking for a location to use as a haunted house attraction for the Halloween season. When he drove past the vacant manor, his first impression was that the place looked like a *haunted house*. Little did he know just how right he was! He contacted the owner and, after some effort, he was finally able to convince him to allow him to open a haunted attraction in the manor.

It did not take long for Matt to realize that some of the occupants from days gone by were still residing in the manor. On one of his first trips there after renting it out, he was doing some work inside the building. He had just placed his saw on the floor next to him after drilling a hole in the wall, when he heard what sounded like a young girl's voice asking him what he was doing. He turned, thinking that his girlfriend Jennifer was playing tricks on him, but no one was there. As Matt turned back, his saw went sliding across the floor. Still not convinced that Jennifer wasn't playing tricks on him, Matt went to look for her. When he reached the door of the manor, he saw that Jennifer was getting something out of the car and was nowhere near the building. Matt was a little spooked at this point, but went back in the house. He was heading upstairs when something shoved him, causing him to fall up the steps. That was enough for Matt for one day, and he left the building.

Matt's girlfriend, Jennifer, has also experienced the strangeness of the manor. About a month after Matt's encounter, Jennifer took several friends to Black Moon Manor who wanted to experience the paranormal activity for themselves. Jennifer told me in an email that some friends and she were in the manor trying to provoke the spirits. According to Jennifer, she was not personally doing the provoking, but was standing right next to someone who was and she did nothing to stop him. At some point, she felt a burning sensation on her backside. Once outside the home, someone checked her back and saw that someone or something had scratched her.

Jennifer believes the reason that she was scratched was that the spirits residing at the manor became upset with her since, as one of the owners, she was in charge yet did nothing to stop the negative provoking that was going on. Since her attack, guests of the manor are no longer allowed to provoke.

By the time this book is published, our investigation of Black Moon Manor will be complete and our evidence reviewed. Will the legends and stories that have been told turn out to be true? Will we experience some of the things that Matt and Jenifer claimed to have experienced, or will we have our own new experiences to add to the already incredible stories? I am looking forward to exploring the haunted side of the Hoosier State and hope to investigate many more mysterious locations in Indiana in the future.

My adventures into the paranormal have only just begun. With each and every investigation and encounter that I have had, the door to the other side seems to swing open a little wider. The wider the door opens, the more incredible and unbelievable things become. When I think I have seen it all, something more incredible happens and it keeps me coming back for more. Sometimes I wonder if it is the plan of the supernatural to lure us back by showing us incredible things.

Since I began investigating the paranormal, I have heard my name whispered out loud. This has happened four times in the last three years. One of my close friends, Lilia, has also heard my name whispered in her ear on two occasions. The first time that she heard the whisper, I had not told her about hearing the whisper myself! On one of these occasions, I was standing at her desk and, in mid-sentence; she stopped to ask me if "I heard that?" When I said, "No, what do you mean?" she explained that someone just whispered my name in her right ear. Every time I have heard the whisper, and every time that Lilia has heard it, it always comes from the right side of our bodies. I told my psychic friend, Cheryl, about the whisper and she told me that it makes complete sense that I hear the whisper in my right ear. She told me that the left side of our brain is the analytical side while the right side is the psychic side of our mind. I have also recorded my name being whispered on several occasions on a digital recorder. It sounds exactly like the whisper that I have physically heard with my own ears. Once you start experiencing these types of things, the world around you becomes a whole new ballgame.

So, if you are not already a paranormal investigator, and decide to become one, to unlock some of the mysteries of the supernatural for yourself, be patient, because the spiritual side is very elusive. But be ready, because as I was once told by my wise psychic friend Cheryl, "Once you open the door to the other side you won't have to worry about traveling to find spirit activity, because they will come and they will find you!" Well, Cheryl was right, and I found out the hard way that what may come to visit you may not be welcome!

ABOUT THE AUTHOR

Larry Wilson spent a decade working as a private investigator, before turning his attention to the paranormal. He is the founder of Urban Paranormal Investigations in central Illinois. He has investigated hundreds of locations throughout the Midwest, and has authored several books on the topic. In addition he has guest lectured on the subject of the paranormal and has also appeared on both television and radio programs.

He is also the founder of 11:11 Films, an independent film company that produces paranormal and historical documentaries. His first film *Strange Williamsburg Hill* is currently being re-filmed and re-edited in High Definition format and should be ready for release in 2015. Larry has also assisted in the filming of three paranormal documentaries for other Independent film companies.

He currently resides in Taylorville with his wife and son.

For more information please visit:

http://lwilsonurbanparanormal.blogspot.com/

Like us on Facebook:

https://www.facebook.com/pages/Chasing-Shadows-by-Larry-Wilson/325540060791693

Photo by Janet Morris

Made in the USA
Lexington, KY
21 March 2018